का

Teaching Environmental Science

Edited & Compiled By

Dr. PREM SUNDER

PRINCIPAL

DARSH COLLEGE OF EDUCATION

GOHANA (SONEPAT), HARYANA

KSK Publishers & Distributors

7/26, Basement, Ansari Road, Darya Ganj,

New Delhi-110002.

Published by:

KSK PUBLISHERS AND DISTRIBUTORS

7/26, BASEMENT, Ansari Road, Darya Ganj, New Delhi-110002.

Ph.: (O) 011-65190343, 43551324, 42564726
(M) 9811088729, (Fax) 42564726, (Res.) 47091605

E-mail: *khelsahitya1@rediffmail.com*

Web: *www.khelsahitya.trade-india.com*

I.S.B.N. - 978-81-89983-20-8

PRINTED IN INDIA 2009

Laser Typeset by:
JAIN MEDIA GRAPHICS
Delhi-110035.

Printed by:
Vishal Kaushik Printers
A-49, Gali No. 6, Jagatpuri Vistar, Delhi-93

Price: Rs. 500/-

Contents

1

ENVIRONMENT—FROM SOCIOLOGICAL POINT OF VIEW

Meaning of Social Issues

Activities of man right from birth to death on this earth are deeply related to social issues. He while living in society cannot detach himself from social issues of life. This is the reason that someone has rightly pointed out that society is the web of social relations and man is social animal. Social issues cover family, associations, institutions and communities. Perhaps family is the strongest agency of social issues. While being in society he has to undertake a number of socio-economic activities such as industry, agriculture, transport, construction of roads, and buildings etc., for surviving in the society.

These activities contribute much to environment degradation including water, soil and air etc. Socio-economic development is the call of the day because everyone desires to have a higher and better standard of living. The geographical land area with man is reducing day by day due to growing population. With the limited availability of land for cultivation purposes, the need for increasing the application of fertilizers, nitrogenous, phosphatic and potash are also necessary for appropriate pest control through pesticides is being made in order to ensure more and more growth of both food and fibre. These socio-economic activities hit environmental conditions. So the protection of the environment is an important issue of the day. There should be development without polluting the environment. Engineers in this respect have a major role to play. Many a time the question comes to our mind as to whether we want environment or development.

To this the straight reply is that, there should be development within the environment. One does not place these two things separate, i.e., development and clean environment. If one cannot breathe pure air, if he cannot drink clean water e.g., if you have come from west and live on mineral water because you do not dare touch water available here, there is something wrong somewhere, and if in the 20th century we have achieved the so called material benefits by the last decade is that we have polluted our land, air, water and food, then there must be something wrong in our achievements we have done for the last 90 years. So the real socio-economic development does not lie in producing and amazing material and non-material things but in having these along with a clean and pollution free environment.

Urban Residential and Commercial Energy Demand

As already explained, the amount of energy required for residential and commercial use varies greatly throughout the world. The economics of energy consumption describe that a country with high GNP use less per capita energy for residential and roughly national use less per capita energy for residential and commercial needs, than a less developed country like Africa and India. 30 percent energy used in North America and 90 percent used in India is for residential and commercial purposes. Reason is very simple. North Americans use air conditioning, refrigeration, water heating and space heating. So 75 percent energy is used in these gadgets. In India, almost all of the energy used in home is for cooking due to scarcity and high cost of fuel. About half of the energy demand in Africa is for cooking because they use an open fire instead of stoves. Using more efficient stoves instead of open fires could reduce these energy requirements by 50 percent. Similarity North America, and India could reduce energy consumption in many ways.

Urban- Industrial Energy Demand- Considerable variation is found in the amount of energy used for industrial processes. Facts speak that 60 percent of all the commercial energy was used in the former Soviet Union. About 30 percent of energy is used by USA for industries.

In the former Soviet Union large deposits of coal, oil and natural gas reserves were available. On account of this reason steel is processed from ore because energy is readily available. Spain and Italy lack large deposits of fossil fuel necessary for the extraction of oil and natural gas. So both Italy and Spain produce steel from scarp steel because it needs less energy. Producing 1 metric ton of steel from iron or in the former Soviet Union uses as much energy as producing a metric ton of steel from scrap in Spain or Italy.

Large capital investment is required to upgrade the process of industrial production, because updated and upgraded machinery and equipment reduce industrial energy consumption. Many countries of the world cannot afford to convert their existing processes due to financial and other constraints. Take the case of India, a nation with few coal deposits, still used the outdated open hearth furnace to produce steel. This process needs double the worldwide energy for producing a metric ton of steel on an average. India being a underdeveloped country if unable to change the process of modern methods has to continue to use energy expensive methods.

Transportation Energy Demand in urban areas. In India energy used in giga Joules per capita in transportation 2,103 in United States, 82 in Australia and 4 in Zimbabwe. These statistics with regard to energy demand in transportation clearly show that energy use for transportation is larger in developing and highly developed countries. Rail, water, private automobiles and mix bus service determine country's energy use for transportation. The places where public transport is efficient less private automobiles are used. This is so because automobiles require about four times more energy per passenger kilometer than bus or rail transport. But in developed countries, the situation is different. The government policy in these countries is different. Government in these countries keep the cost of energy low. Consequently, the automobile plays a dominant role, and public transport is primarily used only in metropolitan areas. Rail and bus transport are about twice as energy efficient as private automobiles. Air travel is relatively expensive in terms of energy.

Electrical Energy Demand in urban areas. Electrical energy consumption in different regions of the world varies widely. The amount of electricity used by all the less developed nations, with over 75 percent of the world's population, is only 66 percent of that used by the United States alone. North America uses electrical energy twenty-five times greater that average per capita use of this energy in less developed countries. The electricity used in industries by developed countries in comparison to underdeveloped countries is also much more.

Unsustainable to sustainable development

Unsustainable development or growth is one in which social issues of life do not progress in a healthy and ethical style. This style of living cover mismanagement of resources like large-scale deforestation, indiscriminate construction and expansion of settlement activities, discharging toxic wastes, pulling air water and what not.

As globalization in its current form increases, so too unsustainable development that it seems to prescribe. There is abudant evidence that the version of modernity demonstrated in current globalization trends neglects scientific evidence and tradition in equal measure. In contrast, the traditional approaches to natural resources management in many communities demonstrate sourced science than development does. Tradition in local resource management has survived over many generations of environmental degradation - which are evidently no longer future threats but very real current crises of - have resulted from approaches to modernity which, in throwing away such traditions aside, continue in denial about scientific evidence.

In many regions of the world, ecosystem stress can be seen due to human-intensive activities such as unsustainable resources extraction and exploitation via large corporations and international trading agreements that open these sources up for excessive extraction and consumerism.

Dropsy is a disease caused due to adulteration in mustard oil by greedy, dishonest elements of society with a motive to get rich at the expense of the lives of the general public in the society. Most of the adulterated mustard oil samples were found containing Argemone, and a few samples were found to be containing adulterants such as polybromide and even mobile oil.

This may be sustainable personal development of an industrialist but most unsustainable development for society as a whole.

All environmental issues involve consideration of ethics. For instance, sufficient food and fabric is available in the world to feed people. It is unethical to allow some people to starve and others consume more than what they need. Developed nations do not realize it. They have enough and still they need more. Their greed for more is ever increasing. They do not surrender before morals and ethics of life and share with others what they have. They must think less about conquering nature and more about learning to work with nature. Each person must realize his interdependence with the rest of the nature including his fellow human beings. To safeguard life on earth, people must learn to control and adjust the balances in nature that are altered by their activities.

One more reason is placed in this context. Members of the society very well realize this view. They view the world's energy situation as serious and have reduced their own consumption. But many do not believe and realize this problem. They continue to consume energy as long as it is available and do not accept any cut on the use of it. Other similar issues are the population problem and pollution. It is ethical to have more than two children when the world is facing problem of overpopulation? Similarly an industrialist may not probably look on the negative aspects of pollution compared to one who is outdoor. Some sixty American cities have been declared having unsatisfactory water supply due to toxicity of pollutants released by industries. Every year in the United States, 20 million tons of paper and 200 million tons of domestic garbage is junked. Keeping this in view the Massachusetts

Institute of Technology, USA reveals that there shall be disastrous collapse in the coming hundred years if both population growth and industrial growth are not brought to an immediate halt. So the first task is to educate economics harpers about the value of environmental ethics. Clean and unpolluted air away from the chimneys of industries, free from fumes and smoke has a great cleansing and purifying effect for sustainable growth. For this purpose, taking ethical stand is important regardless to those who disagree with it.

Emerson, a pioneer philosopher environmentalist in his first essay Nature published in 1836 objected fast economic development by way of invading nature. Another naturalist who held the beliefs similar to Emerson was Henry David Thoreau. According to Thoreu, "Truth in nature and wilderness over the deceits of urban civilization." He summarized his feelings about nature as But most men, it seems to me, do not care for nature and would sell their share in all her beauty, as long as they may live, for a stated sum-many for a glass of rum. He further concluded Thank God, men cannot as yet fly, and lay waste in the sky as well as the earth.

Urban Problems related to energy

Energy is a essential of a man where-ever he lives-urban or rural society. In urban areas the need of energy is increasing by leaps and bounds. Take the case of mega-urban dwelling places. Vehicles both petrol and engine driven are seen flowing like floods. According to one rough estimate nearly 40 lakh surface transport move on Delhi roads. These vehicles run with the basic input called oil which is one form of energy. What about other sources of energy-renewable and non-renewable being used by urban population. The world-wide demand for energy will be more than double by the year

2020.

Moreover, countries use energy in an uneven manner in the world. Some countries use more energy while many less. In developed countries the amount of energy used is much more compared to underdeveloped developing countries. Industrialized developed countries use energy for these purposes:)i) residential and commercial, (ii) industrial and (iii) transportation. In less-developed countries, most of the energy is used by individuals. Canada and United States use about twice as much energy as someone in France or Japan and over twenty-five times as much energy as a person in Africa. The consumption of energy in India is more than Africa because it is the second populous country in the world with population touching slightly more than 100 crores, according to the latest census.

In India, about 70% commercial energy is being generated through coalfired thermal plants. Coal consumption by thermal plants is nearly 130 million tonnes annually and is expected to rise to 272 million tonnes by 2002. There are about 40 thermal plants both big and small scattered all over the country. The Korba-Super Thermal Power Station is the largest. It has installed capacity of 2,100 MW. Power from the project goes into the western grid comprising the states of Madhya Pradesh, Maharashtra, Gujarat and Goa. However, Bihar, West Bengal, Delhi, Gujarat and Haryana are some of the main producers of thermal power in India.

Sustainable Development

Continuous and persistent rise in any system is known as sustainable growth. It means continuous rise in the GNP of the country resulting into the rising trend in per capita income of the country. The objective of sustainable development is to promote development which in turn reduces the disparities in life styles. Improve global consumption and maintaining a healthful local environmental and thereby contributing towards solving critical global environmental management problem:

1. Reduce disparity in life style.
2. Protecting soil from pollution.
3. Reduction in waste and prevention of pollution.
4. Integrated environmental systems management.
5. Efficient use of natural resources.

The 10th FYP envisages the provision of potable drinking water to every settlement in the country on a sustainable basis and the pursuit of all possible measures for the rapid expansion and improvement of sanitation facilities in rural and urban areas.

Water conservation, Waterland management and water-harvesting

Water conservation means saving water for future use. When water scarcity is felt storing water and avoiding wastage of water is equally important approach towards water conservation. Wetlands are found of course all over the country and are either natural or built by people. Over the years they have gradually depleted, leading to a number of problems such as flooding, water scarcity, and water logging. Keeping this problem in view, takes have been constructed in the catchment areas of cities that were not located near a river or a large lake. Rainwater or run-off would collect during the monsoons and be stored for the rest of the year in these tanks and lakes. The city would get its water supply from these.

In southern India, the towns were built around a temple, which always had a tank at the centre. Wells were located in the tank, which served as a link with the aquiferous. There were innumerable tanks dug by rules, most of which have been destroyed. In fact, in some areas tank building by the ruler was considered a noble deed.

In Karnataka, the Hoysalas built tanks all over the state. Some of these still exist. Festivals were organized around tanks; cleaning and de-silting the tanks were a part of the rituals. But over the years, human activity steadily destroyed these wetlands. In Bangalore, the city bus terminus has been built on the Dharmambudhi tank. A part of the Sampangi tank has given way to the Kanteerva stadium and the remaining to a housing colony. Another lake was filled up to house the city market.

The Cholas were well known for the construction of tanks in the state of Tamil Nadu. But today, despite heavy rains, Chennai reels under severe water crisis along with flooding. A number of housing colonies have been built on tanks and over the years the numbers are increasing. This is the case in most large cities in the country.

Energy problems of urbanized population

The details of energy demand clearly reflect that there is imbalance between demand and supply of energy. The demand of energy is more whereas supply is less. Demand is more because number of houses and vehicles are increasing day by day. Besides this, consumers of energy are abudant in number. We were 34-37 crores in 1951 but now we are more than 100 crores in India. Every one nearly is having electrical appliance. The energy consumption is therefore shooting up. On account these reasons urban people are feeling following energy demand problems.

1. Where you could walk, walk or at the most use a bicycle.

2. Eat grain based diets than carnivory.

3. Renewable energy damages are less in comparison to coal and oil. Use it frequently.

4. Create wind-break on the north side by house and plant deciduous trees on the south side.

5. Keep an eye on energy use habits. Turn-off lights, televisions, geysers, computers etc., when not in use.

6. Careful use of large electric or gas appliances such as stoves, refrigerators, washing machines and dryers.

7. To have sustainable life-styles, use of lesser voltage sources of light may be exercised to conserve energy so that even distribution of power is done.

8. Hydro-electricity is not sufficient and where there are no water flowing or perennial rivers, there were cannot depend upon hydro-electricity either.

9. Entropy is failure because eventual heat is not distributed within a closed system.

10. Transmission lines are not perfectly erected.

The Wetlands

Wetlands are areas lying along the banks of rivers and lakes and the coastal regions. They are life-supporting systems providing fish, forest products, water, flood control, erosion buffering, a plant gene pool, wildlife, recreation, and tourism areas. Though they are endowed with a rich biodiversity, yet of late they are being greatly exploited. Wetland come in many forms. Swamps are good example.

In most simple manner wetlands represented interface between land and water. Depth of wetland should be a maximum of 6 metres according to the Ramsar convention. For minimum depth there is no limitation, it can be 6 inches or no water at all in dry season. Wetland are important links between land and water. They are regarded as most important ecosystems on the earth, and called kidneys of the biosphere as they moderate flow of nutrients and silt from land to water by trapping them. About 6% of the land surface of earth is wetland. Many wetland species have become threatened and endangered because of their dependence on a particular type of wetland ecosystem, which has become seriously degraded or destroyed.

Functions of wetland

These high productive ecosystems have important benefits such as:

(i) maintenance of integrity of other ecosystem.

(ii) climate stabilization.

(iii) toxicant removal by doing water treatment.

(iv) salinity control.

(v) erosion control.

(vi) ground water recharge.

Types of wetlands

Wetlands can classified into three categories:

(i) Marshers, (ii) Swamps and (iii) Peatlands. Within each of these categories wetlands can vary widely, because wetlands depend on water resources, their boundaries can change. The characteristics that describe each category include vegetation, soil type, water supply and water chemistry.

Wetlands in India and Threats to Indian Wetlands

In India wetlands are distributed in different geographical regions. As per a survey conducted by ministry of environment and forests, the total wetland area of India is about 4.1 million hectares and the rest is occupied by man-made wetlands. Maximum number of natural wetlands are in Assam. Total no. natural wetlands in country is about 2,200. Main threats to Indian wetlands are:

(i) Use of poisons and explosives during hunting and fishing.

(ii) Grazing by domestic livestocks.

(iii) Degradation of waterlands, increased siltation of wetlands, deforestation and soil erosion.

(iv) Encroachments for road and other transportation facilities.

(v) Infestation with aquatic weeds uncontrolled growth of weeds and eutrophication.

(vi) Cutting of timber, trees and mangroves for fuelwood and domestic use.

(vii) Alteration of wetlands to other use categories such as for agriculture urban development, industrial sites, forestry operations etc.

(viii) Pollution from waterboats, domestic sewage, pesticides, fertilizers and effluents.

Environmental Ethics

Earth is the most beautiful object. It is unique among all the planets in our solar system. It is a pathway of sustainable development. We abuse land because we regard it as a commodity belonging to us. Once we see land as a community to which we belong, we may being to use it with love and respect. Environmental ethics teach us value of earth; interdependence; belongingness and ecological preservation-necessary for sustainable development. The treatment given to earth in the form of commodity is most unethical recognized by ethical environmentalists. This is so because many people see little value in an undeveloped river. They feel it is unreasonable to leave a river flowing in a natural state. This is the reason that many rivers throughout the world have been controlled to provide power, irrigation and navigation for the people at the expense of the natural world. Sarovar over Narmada and Rajasthan Canal Project are a few important river valley projects constructed by man for power generation, navigational, recreational and irrigational benefits. Power a most versatile form of energy is being used in the development of agriculture and industry.

To meet the ever increasing demand of electricity, public leaders are engaged in harnessing natural resources amongst which water is one. Not using such resources would be wasteful is a feeling quite common in their minds. A serious conflict is going on over the value of old-growth forests in the Pacific Northwest of the United States. Many economic interests are involved over this dispute. They want to use the forests for timber production. If not to make their use for the betterment of society, they feel. Those who support land as a community to which they belong to not agree with this attempt. They feel that all the living things that make up the forest have a definite value which we do not appreciate. Felling and removing the trees would destroy something that took much more time to develop and may never be replaced.

120 nations at a global meeting on CITES objected the trade sanctions against china and Taiwan when these two nations were found selling rhinoceros and tiger parts for medicinal purposes. The animals are killed to provide material for growing international market. Rhino horns and tiger bones are still used in traditional

Chinese medicines to treat fevers and joint as well as bone are available for sale in pharmaceutical shops as well as government run stores in China. This may be one important reason that fewer tigers and rhinos remain in the world. In view of this reality. CITES has proposed a cut on international trade in endanger species. Corporate leaders should support in leaving aside economic interests.

Though there may be many environmental attitudes yet for the sake of convenience three classifications are made and each one in labelled separately provided with suitable heading.

1. Conservation ethic - This is the ethic of environment. Conservation means maintenance and preservation of the environment to meet human needs for production while insuring that proper consideration is given to aesthetics and recreation. The conservation ethic is related to scientific preservation. Through it recognizes the desirability of decent living standard yet works toward a balance of resource use and resource availability. The reasonable use of Earth's natural resources such as water, soil, wild-life, forests and minerals is the major objective of conservation ethicists. The conservation ethic suggest a serious balance between total development and absolute preservation of resources.

An effective conservation programme results in continuous production of plants and animals and mineral resources which can create imbalance between availability and demand of such resources may not today but in future definitely. The goal and objective of conservation ethic is one people living together in one world indefinitely. Conservation is therefor very important in view of this fact. Environmental ethicist can better guide human beings over this issue and preach. As humanists and populists, they consider the goals of conservation to prevent waste and save resources for prudent constructive and efficient human use and absolute preservation of resources. Social ecologists suggest that the problem of human ignorance and greed can be solved by interacting with correct thinking people for the survival of planet. Canada's Green Plan developed in 1990 address current environmental problems. It contains comprehensive new programmes, policies and include: (i) Protecting and enhancing the quality of water resources and promoting wise and efficient use of water, (ii) Developing an environmentally healthy society, pollution problems did arise. Groundwater became contaminated with chloride from the drilling. Forest's most prized commodity solitude is lost. The case of the Pigeon River Forest is solitude versus oil. Agreed that a barrel of oil has direct economic value. But what value we can place on a day of solitude in nature is difficult to estimate. (iii) Cutting waste 50 percent by the year 2001. and (iv) Securing current and future generations a safe and healthy environment and a sound and prosperous economy.

The question arises what is enough, Many poor societies have little but do the rich societies agree to halt. The developed societies are over-ambitious. One thinker said The earth provides enough to satisfy every person's need, but not every person's greed. Contentment is great thing but bow many nations compromise over it. Growth and domination is the central sociocultural goal of most advance societies whereas economic growth and rcsourcc exploitation are attitudes shared by developing societies amongst which Indian society is one. We continue to consume natural resources with this feeling in mind that supply of resources will never end. This is reflected in our increasingly unstable relationship with the environment. Our desire for taking resources is growing day by day from the common lot regardless to future.

2. Preservation ethic - This ethic reserve special place for nature. Different preservationists have different reasons to preserve nature. Some have religious reasons while many shoe recreational aesthetic feelings towards preservation of nature. Those who have religious belief towards nature hold reverence for trees and respect the right of all creatures to live. To these preservationists social and economic costs are not so important. Preservationists opine that when a tree is hexed it may not show any visible feelings of suffering but that does not man that we can go to felling trees. Aesthenic or recreational preservationists believe that nature is beautiful and refreshing. It should be available cross-country skiing, picnics, hiking, camping and just for quietness and peace. Exploitation should be curbed. Glorious pursuits beyond sensible level of need must not be permitted. With these feelings, man can live more happily in the future because harmony between him and nature is created by way of eco-friendship.

The views of semireligious and recreational preservationists seems to be essentially scientific. These individual argue that human species depends on and has much to learn from nature. Natural resources such as trees and plants are the life and blood of human civilization. They have served human species in various ways from the days of yore. A large portion of the food, medicines and materials that man consuls is derived from some wild species of plants Catharanthus rosens is one such plant which originated in Madagascar when used in the form of herbal medicine, treated children suffering from Leukemia. Dr. Richard Evans rightly pointed out The drugs of future grow in primeval jungle. Man should know the immense value of nature by putting aside ignorance and shortsightedness. This ignorance has put about 25,000 flowering species on the verge of extinction. Preservation of nature alone can save human species. Preservationists comment and argue that rare and endangered species and ecosystems, as well as, more common one must be preserved because of practical utility. Remember natural diversity and wilderness is much more superior to humanized domesticity. If we care for Nature, it can be rich, bountiful and inhausibly sustainable.

3. Progressive ethic - It is also called development ethic. This ethic is governed and based on action. Progressive ethic reflect that human race is master of nature. Earth and its resources exist for the benefit and pleasure of man. Who denies than man should not make use of natural resources. It is called work of ethic which dictates that human should be busy creating continual change. The activities performed by him that are bigger, better and faster represent progress itself is good. This philosophy is encouraged by the idea that if it can be done, it should be done. Our actions and energies should be best harnessed in creative work. The dream of upward mobility is embodied in development ethic. The notion that bigger is better, definitely not new to us but is unpatriotic when one hit the environmental ethic. Preserve ethic and exercise growth is the answer.

Urban Wetlands

There are no rivers in the vicnity of Bhopal and the city depends entirely on lakes to meet the water needs of the city. Over the years, these lakes have become polluted due to the dumping of sewage into them. The government recently announced the Bhuj Wetland Project, a scheme to prevent the flow of liquid and solid waste into the lakes and also to desalt them. More than 2000 ha of land around the upper lakes, which feed water to the lower ones, is to come under plantation.

Once upon a time, these urban wetlands maintained a steady supply of water of the city, recharged the groundwater, cooled the city, and prevented flood as they have natural drainage system. Over the years, these very tanks and wetlands have been neglected, encroached upon either to accommodate more houses or to dump waste. This has caused large-scale water crisis and monsoon flooding and water logging, leading to misery and disease. These wetlands that have been protected for centuries are now being ravaged and destroyed in the name of development.

People in the urban areas learn to live in harmony with nature in their own habitat. The wetlands, the green belts, and the flora and fauna of these areas have to be preserved for urban areas to survive and remain healthy. Many people do not understand the value of wetlands, in support we give following cases.

1. Indore's water needs were met by a large numbers of wells and tanks located all around the city. But over the last few years, the tanks have gradually dried up, including the largest of them, the Yashwant Sagar Lake. Vegetable and fruit cultivation along the banks of the lake has caused pollution due to run-offs from the fields. This has caused eutrophication and contamination.

2. In Kolkata wetlands, both artificial and natural, have been a part of the city system. The main drainage for the city has been the Hooghly River; but the incline of the city is eastwards and this has led to the creation

of the East Kolkata Wetlands. The Salt Lake is one of the largest wetlands in the city and serves as a flood cushion for the city. Kolkata has no sewage treatment plant. The sewage is treated in the sewage-fed fisheries purify this water through a series of sedimentation tanks and use of the water for fish cultivation and agriculture.

Corporate Environmental Ethics

Corporation also called a corporate enterprise is a large business house managed by professional managers, designed to operate at a profit. Corporation have legal entities. Examples of corporate enterprises in India are D.C.M., Modi, Reliance and ABB. A corporate enterprise procure raw materials, manufacture and market the products so produced. These corporate enterprise when seen in the shape of industry appear dirty because all industries consume energy and resources. Some waste is inevitable when raw materials are processed. Some of the waste material may be simply useless. For instance, the food service industry use energy in the preparation of meals. Much of this energy is lost as waste heat. In addition smoke and toxic gases are released into the atmosphere. The cost of controlling waste can be very important in determining the company's profit.

Ethics are involved when a corporation do not maintain standard as prescribed of the product and waste disposal to maximize profit. Ethics appear when an industry cut corners in production quality and many times present adulterated items to the users. Reduction in cost in another issue before corporation. The cheaper it is to produce an item, the greater the possible profit. It reduces the cost of production by lowering the quality standards which is highly unethical. To an industry it is cheaper to dump wastes into a river than to instal wastewater treatment facility, and it is still cheaper to release wastes into the air than it is to trap them in filters. Pollution in this way released is considered immoral and unethical to many but to the corporations, it is just one of the factors that determines the profitability. This is true in chemical, oil and many other industries. If we question corporate leaders over this subject then the simple answer given is that stockholders expect a quick return on their investment. It is, therefore, observed that corporations often make decisions before short-term profitability rather than long-term benefit to society. It is not an unethical practice exercised by corporations.

The expansion of a corporation depends upon the amount of profit it realizers from a given investment venture. Higher profitability leads to higher expansion. Take the case of big corporation like Tata, Tata, Birla, Dalmia and IBM Corporation. Analyser predict more expansion of these organizations in future, because these corporations create increased demand fort their products through advertisement, publicity and other marketing strategies. The more they expand, the more power they attain and greatly influence decision makers who create conditions favourable to their expansion plans. So it becomes a never-ending spiral. Who stops their expansion provided these corporations do not show irresponsibility toward the environment. The corporate decision maker exercise minimum discount to the public interest, while maximum efforts are made to increase profitability. It is fair to enhance profitability by increasing social costs. One is amazed to notice that many political and legal institutions defend private property rights rather than social and environmental concerns.

To evaluate corporate environmental responsibility

Valdez Principles were coined as a result of the 1989 oil spill in Alaska. These principles were later adopted by the CERES organization. These are recognized as CERES Principles propagated by independent group of environmentalists and social investors. These principles are set of codes against which the working of business houses as well as corporations can be assessed and compared. The principles encompass making sustainable use of renewable resources by way of minimizing pollutants. Corporation are expected to make compliance with CERES principles for doing business. The need for developing such principles was felt due to deaths of ducks that occurred on account of oil spill in Alaska in 1989. There is no sanity to maintain higher industrial productivity at the cost of depriving a nation of breathable air and drinkable water, is clearly reflected by CERES principles. CERES principle as its core advocate environmental justice which means protecting health of human beings and the productivity of ecological systems upon which all human activities rest.

Societal Environmental Ethics

Societal environmental ethics concentrate on the development of environmentally literate society. Accomplishment of it is a difficult task because society is composed of a great variety people with different views and opinions. However, this variety can be distilled into a set of ideas and observed from the ethical point of view. Western developed societies accommodate unchecked growth and development because members of these societies accommodate unchecked growth and development because members of these societies feel and act that earth has unlimited reserves of natural resources and limitless ability to assimilate wastes.

Continual growth is the rationale of developed nations behind this thinking. Unfortunately the idea of continual growth has not always been carefully planned and implemented. This growth has resulted in the use of nonrenewable resources for comfortable homes and convenient transportation. Oil was discovered by man in a remote area of Michigan known as the Pigeon Reiver Forest in 1968. Its 37,600 hectares is considered to have all wilderness qualities because many animals inhabit this region. Oil wells were drilled in the area with a view to overpower energy crisis. Michigan would get several hundred million dollars of royalty against this venture. But what happened, much less oil was found beneath the forest than was originally predicted with all safeguards. This attitude of society towards environment is deeply rooted since time immemorial. The first settlers when arrived in North America considered nature an enemy. They viewed nature to be conquered and vanquished. Invading and exploiting wilderness were justified by religious beliefs by them. This attitude towards nature is still popular amongst them. Not developing a resource is considered almost a sin by many.

Pollution Abatement Eco-friendly

The chambers of Commerce and Industry Association have an important role in sharing the experiences of industrial units besides providing a form for exchange of ideas and technical knowledge in the implementation of pollution control schemes. A survey conducted by ASSIOCHAM revealed that one of the basic problems faced by the industry in implementing pollution control system is that proper monitoring devices are not available within the country. Based on the survey ASSOCHAM has submitted a list of monitoring devices which can be manufactured in India.

Various interactions between the industry and Government organised by ASSOCHAM have highlighted the following suggestions for effective management of pollution:

1. In setting standards, parallel supporting system and facilities for effective implementation to be given equal attention.

2. Creation of centralized data bank pooling all information on wastes particularly hazardous types.

3. Greater coordination between industrial units and multinational projects being set up in India with latest technology in this area.

4. One or more environment performance units can be set up to tackle the problems of pollution in the industrial sector enabling the other units to draw upon the expertise of these companies.

5. Many of the existing standards are unachievable for the existing units because of the type of technology in use, age of the unit, size of the plant, lack of economically viable pollution control technology, non-availability of testing equipment and non-identification of dumping sites for storage of wastes. In view of these considerations, there is need for evolving rational standards and different norms for new units and the old ones. Besides standards need not be the same as in developed countries.

6. The role of government enforcement agencies should not be restricted to only regulation but should also provide information education, guidance and counselling.

7. Removal of prejudices against recycled products through education campaigns and providing offtake guarantees by institutional buyers to promote waste

utilization for manufacturing useful products.

8. The proposed institute could also industry particularly in the small sector through constancy, training and providing technical assistance.

9. An autonomous industrial waste management institute as proposed by National Waste management Council should be set up to provide institutional support to the industry. The institute could also advise the Government on policy issues and evolve strategies for reducing waste and in managing it.

10. A national policy needs to be evolved on the management of hazardous and other wastes. It should aim at reducing the volume of waste, utilizing it for manufacturing recycled products to generate resources, treating and disposing the residual waste in an environmentally friendly manner.

Now standards for bottle water should be adopted. New BIS code says packaged water will have to be tested for 32 pesticides; individually, the quality of those pesticides cannot exceed 0.1 micro-milligram per litre and total pesticide residue should be less than 0.5 micro-milligram per litre. Licensed manufacturer shall have to obtain permission from Central Ground Water Authority, because most bottlers use laced with pesticide in most areas. Small companies will be wiped out because new plants could cost over Rs. 15 crores. Instead of burning the leaves, bury them in compost pits, by which it gets converted to natural manure, making it beneficial for the soil.

Economic aspects of environment

Many people hold the view that there is conflict between quality of the environment and economic growth. Working out environmental benefits quantitatively is a difficult task. Though economic growth depends upon environmental conditions yet it is difficult to work out cost benefit analysis between the two. Attempts have been made to estimate monetary losses and gains as a useful tool in the formulation of development objectives and in the execution of environmental policies.

Costs

A study revise that the effect of noise by London Airport on the value of property nearby and on the elimination of noise did not make people conscious about the monetary benefit represented. On the contrary the occupants of the expensive houses gave higher value to noise elimination while poor residents attached more value to being close to the place of work. The possible effects on the future generation whose voice is not heard make some of these views or estimates mor valuable.

Of all the various environmental factors which have economical consequences, the most well known and easy to identify and analyze is that of the industrial pollution and urban wastes. The economics of industrial pollution, which is basically in terms of effluents, wastes and residues, have been studied in many countries including Japan, Italy and United States. The economics have advocated that polluter must pay for the pollution. There are also other views that the industrial polluters pay charges which are directly proportional to the amounts of wastes or effluents discharged, as in the polluter pays principle used by Environment Protection Agency of the USA.

Scaled compensation in another economic method and aspect of environment. Though technically not possible, yet it has been introduced in Japan and Italy. The payment of compensation has been enforced by law courts under various environmental and civil laws. Under this, people are compensated if they are their property suffer from environmentally bad conditions.

One of the unresolved issues today is to what extent is the ecosystem and our environment going to remain stable under the stress of various development actions. Scientific research can throw some light on this issue in the future. Techno-economic analysis of the nature of developmental alternatives is also of crucial importance, to recognize the problem of harmonizing economic and environmental conflicts. Human settlements, industrial and agricultural activities lead to production of wastes of various kinds. The overall term generally used for these, are residues. If we use the term resume broadly then,

1. One part of the residues, again termed generally waste, can be completely foreign to nature and may represent a serious source of trouble for living organisms and the equilibrium of the ecological system into which these wastes are put. The polluting effect of a residue, in whatever category, has to be taken into consideration both the nature and the quantity of the waste discharged and the dimensions and capacity of the receiving system to absorb and transform it. Included in the concept of residues and wastes is also heat and this is termed as thermal pollution because a rise of temperature by even a few degrees in a lake or a river may change the ecological system and change the species composition drastically.

2. A part of residues which can be termed waste is biologically or geochemically degradable and at the end is converted into naturally occurring substances such a nitrogen, carbon dioxide, water, metal oxides and others. However, the process of natural degradation takes placed at the expense of the quality of the environmental

and, therefore, has an impact on human society. For instance, a human or animal excrement discharged in the water is degraded in lake or river but in that process takes up oxygen from the river or lake and decrease the capacity of these waters to receive further bio-degradable wastes. In fact, it may even jeopardize fresh water fauna through lack of dissolved oxygen.

3. A part of the natural resources is definitely subtracted from the reserves when they are used in buildings, roads and other semi permanent structure.

4. The amount of residues tends to be greater than or equal to the amount of natural resources subtracted from the reserves; because in the use of resources, in the manufacture of commodities or in the combustion, oxygen or water may be subtracted from the atmosphere or river and converted into products such as carbondioxide and water vapour which are also essential residues but which cycle quite easily in nature.

The number of pollutants are so many that it is often difficult to study the natural history of a single pollutant. There are also complicating possibilities of synergistic effects when two or m ore pollutants are present together. The economic effects in terms of costs of remedying or of prevention are difficult even to estimate. Nonetheless, there are costs or sacrifices involved in the process of absorption of residues even though they may not be toxic.

Handling of agricultural residues or wastes is comparatively easy, except for weedicides and pesticides, etc. The techno-economics of agricultural residues, however, provides exciting possibilities and challenges. The potential of utility of these residues can extend to the stage of complete recycling in an energy and ecologically efficient manner. Animal wastes or residues are also generally treated as part of agricultural wastes. However, the chemical composition of animal wastes can vary greatly depending on the nature of the waste and the manner in which these are collected and used. These wastes are a resource regular use by our people in a variety of ways but are often lost sight of. In fact, there is substantial resources value in animal residues which can be increased by suitable categorization and processing. When we use plant wastes as fuels it is necessary to make cases the gain may be significant. With the development of controlled microbial degradation, the potentiality of getting fuel as well as recovering a part of the nitrogen and phosphorus for recycling as fertilizers makes such economic calculations worthwhile under different assumptions and various models.

Similarly, on the negative side, for instance, the use of lead in petrol as an additive does not involve only an immediate hazard to man. Lead which settles on the roadside vegetations and soil can get into the cattle and indirectly into the human system. The economics of lead additive is not only the gain in improved combustion of petrol engines but the cumulative costs of lead in the vegetation of the neighborhood which has a substantial probability of getting into human and animal systems of inhalation of getting into human and animal systems of inhalation or through the food chain. The statement that the final destination of lead in soil or water does not give us any indication of the possibility of damage to human or animal health or the disturbance of biological equilibria in the process. Similar situations or comparable statements can arise about several other heavy metals. However for operational purposes, the pathways, alternatives, costs of amelioration have to be considered in the quantitative manner. Such studies are needed to evaluate technological and economic factors in the disposal and reuse of wastes.

One of the concepts often used to quantify some of the environment benefits is the concept of augmented gross national product. The AGNP is defined as the GNP plus the value of all other public golds and non-market services. If the industries on the banks of the Hooghly at Kolkata run efficiently to maximum production, then they are points on the production possibility frontier. Suppose we want a river-front park; it can reduce the production possibility by making less of river bank available for industrial use. However, efficiently run godowns or industries could give up some river front areas by reorientation and better use of land without losing access of the river.

Practical problems such as the choice of location of a new plant or the choice of a technology or a choice of an anti-pollution policy have financial effects as well as long-item economic and social effects. More accurate information is usually necessary on the entire cycle, energy needs and the types and amounts of various residues, to be able to quantify the increased benefits derived by taking environmental effects into consideration. The possible effects of residues on geographical, geochemical and biological conditions of the receiving bodies are necessary to determine such choices and to study comparative risks and costs over a longer time profile. Very often it may boil down to differing short and long term views. The estimated span of time during which a process or an industry will operate is used to calculate future residue bonds, due to explosion, or starting of new industries and possible changes of technology in existing industry.

Finally an important and developing concept is that

of considering residues are resources rather than wastes. For instance each market in Kolkata produces several forms of vegetable, animal and other residues. The management of urban solid wastes now being discussed by economists and technologists is utilization of organic wastes after separation of paper, metals, glass and plastics. Today these organic wastes are often used as landfills. However, they can be microbial degraded to obtain methane for fuel and the residues usually rich in nitrogen and phosphorus can be used as fertilizer. Sometimes some of the micronutrients are also present in such residues and improve fertilization of the soil. Human and animal excreta can also be similarly processed. Thus there are both economic and environmental gains in such approaches. However, changing costs ad improvements in technology will increase the attractiveness of the approach.

There are some problems with plastics, pesticides and other non-biodegradable mater when these are mixed with such wastes. Most of the plastics are not easily biodegradable and some plastics such as PVC cause problems by inhibiting microbial degradation. PVC an lead to the production of chlorine which is toxic, or hydrochloric acid on incineration. It is often recommended that the glass, paper, plastics and metals be separated from the rest of the residues as all these are amenable to reuses through appropriate processing. However, the degree of social organization necessary for such segregation and recycling has been difficult to achieve till now. This has made the problem at present unsolvable in our society. To some extent, however, glass, paper and metal is recycled because of small traders who buy waste paper, cans and glass from houses and sell to the reprocessing industries.

The problem of recycling also has to deal with the total energy and social costs in the process of segregation and recycling. In many countries, and this is particularly true of India, it appears that due to the rising costs of energy and extracting and processing minerals, recycling of some wastes of industrial processes and domestic use might become increasingly economic in the future.

Role of Science and Technology in Development

Interrelationship between Socio-economic and Scientific Factors : Scientific factors and socio-economic conditions are closely related to each other. As a cause, economic factors tries to know the level and type of technological needs, ability of society to procure them, resources availability for research and development and efforts and capacity to build scientific and technological institutions. There is no doubt that capital intensive technique ensure higher rate of production quantitatively and qualitatively but poor countries are not able to buy these due to a number of reasons. They lack scientific and technical manpower too. On account of this reason they do not have good science and technical colleges. They do not have sufficient resources for building industries and certainly nor for research and development. Similarly social conditions such as values, attitudes and motivations are necessary for promotion of science. Traditional societies are quite backward having no scientific temperament. In India political leaders, intellectuals and practicing scientists have often stressed the need to develop scientific temperament among the people which they find to be lacking due to various historical, ethnic and ethical forces active in society for centuries together.

It is, therefore agreed that science and technology

play a very crucial role in development of a country. The successive five year plans have stressed the need towards scientific and technological maturity. Per capita income and standard for living depend heavily on technology. Some significant contributions of science and technology towards development are stated below:

1. Development of innovation and discoveries.
2. Employment generation and removal of poverty.
3. Reduction and regional imbalance.
4. Development and growth of human resources including food, cloth, health, sanitation, hygiene, nutrition and water supply.
5. Conservation of resources.
6. Exploration and extraction of new natural resources.
7. Advancement and improvement in defence capabilities.
8. Increased productivity qualitatively and quantitatively in both agricultural and industrial sector.

Methods used by engineers for controlling environment and ecology

Two methods viz:

1. Cleaning gaseous effluents - Removal of gaseous pollutant emissions is important as complete source correction is not possible in existing units of production. It is cheap and economical too if pollution control equipments are installed for removing pollutants from the main gas stream. Such emission control equipment are classified into two categories. These are: Equipment for particulate control and Equipment for gases control. Particulate matter emitted in gaseous streams range in size from 100m down to 0.1m and even less. Following

types of equipment are used for removing particulate matter from gas streams.

(i) Elecostatic precipitators - It is process in which particles suspended, get separated in gas stream. These equipments are mostly used in powerplant, cement, paper mills and oil refineries etc.

(ii) Fabric filters - It is one of the efficient methods for removing particulate matter from industrial gases.

(iii) Cyclone operators - It uses centrifugal force generated by a spinning gas stream to separate the particulate matter from the carrier gas.

(iv) Gravitational settling chambers - It is a simple type of equipment used in gaseous stream with its entrained particles allowed to flow at a low velocity in the settling chamber. It provides sufficient time for particles to settle down.

2. Source Correction Method - The best way of controlling air pollution is preventing pollutants or minimizing the pollutants while planning the erection of an industry at the very stage of process and design. In an existing industrial units source correction is default. However some correction can always be applied in the industries which have already been installed by raw material changes, modification of process equipment and process changes. These are explained below:

(i) Process changes - Process changes involve modified or new techniques which lower atmospheric pollutant emission. For instance if we wash coal before pulverization then it can reduce fly ash emission.

(ii) Modification of process equipment - Emission of pollutants can be minimized by replacing process equipment. For instance unburned carbon monoxide and hydro-carbons can be made to burn fully by suitable modification in the carburation of automobile engines.

(iii) Raw material changes - It has been established that raw material is responsible for causing air pollution. Use pure and A grade raw material. It offers best solution as it reduces formation of undesirable impurities.

Role of Engineers in Environmental Protection

People are demanding as a fundamental right, not only clean air and pure water and food, but also decent housing, an unpolluted land, freedom from excessive noise, adequate recreation facilities, and housing in communities that provide comfort, privacy, and essential services. The engineers and planners must ensure at the planning stage itself that the projects are so designed as to pleasing and enhancing the well being of the community. Before any project is undertaken, a site is selected wherein following considerations are important:

1. In relation to occupational health-safety, toilet, shower and locker facilities, lunch room, clean air, safe water, accident and explosion hazards, pathogens and dust in work areas ventilation, general sanitation and compliance with the Occupational Safety and Health Act should be ensured.

2. Engineers should see that permits are obtained in compliance with the clean Air Act.

3. Solid waste disposal should be in compliance with Solid Waste Disposal Act and Resources Conservation and Recovery Act.

4. There should be perfect provision and arrangements with regard to waste water disposal from dumping platform, storage and charging areas, gas cooling, toilet, shower, dressing rooms, waste-water reuse and cross-connection control. All measures

adopted should be in compliance with Clean Water Act.

5. Planning of water supply should be done for drinking and sanitary purposes, for cooling water, for make up water, and cross connection control. Water resources, recreation and wildlife. There should be compliance with Safe Drinking Water Act.

6. Noise level should be controlled, hopper charging and truck delivery and leaving should be convenient.

7. Special care should be taken in matters of drainage and flooding. The site selected should be above the 100-year flood level before erection for the structure is started. Access to roads, well drained and above flood level, is equally important.

8. The site selected should be good from the geological point of view. It should have suitable foundation soils, and be safe from earthquakes. The use of limestone, dolomite or gypsum should be avoided.

Wild life Protection: Legislative Measures

Undomesticated animals living in their natural habitats such as forests, deserts and grasslands etc. refer to wildlife in layman's language. To an ecologists, the meaning of wild-life is somewhat different. An ecologist includes both fauna naturally occuring animals as well as flora plants in wild-life. Wild-life, endangered and super species are extincting from the lap of the nature.

Indian Board for Wildlife : It is the main advisory body to Govt. of India. At its first meeting, the Board made a recommendation for unified legislation for wildlife conservation in India. The wildlife Act was enacted in 1972 which has been adopted by all states. Biological diversity in an important and irreplaceable asset. We are adopting national legislation and international treaties to protect endangered species and biodiversity. Following steps have been taken in this direction:

1. The Biological Diversity Bill 2000, awaiting Parliamentary sanction and presidential approval, is a landmark legislation aimed at protecting India's bio-assets and ensuring equitable benefit-sharing by the users and the owners.

2. U.S., Endangered species Act of 1973 and COSEWIC 1976 provide powerful approaches to wildlife protection.

3. The endangered species Act and CITES represent a new attitude towards wildlife in which we protect organisms just because they are rared and endangered.

4. Worldlife protected area network comprise 84 national parks and 447 wildlife sanctuaries.

5. Recognition for Ecological conservation.

Consumerism and Waste-Products : Environmental protection Act

Consumerism is an activity performed by consumer in purchasing goods and services. Waste-products are residual left after consumption having no utility to consumer and thrown or put in waste pots also called litters as well as Garbage bin. The development of a nation can be measured in terms of per capita generation of garbage. The figure ranges from 0.5 kg. per head per-day in developing countries to 2.5 kg in advanced countries.

Waste Products are horrible because insects and rodents proliferate in garbage and transmit plague, typhoid, cholera, skin diseases and even malaria and filaria with the breeding of mosquitoes. Garbage dumps contribute to the seepage of toxic metals and other chemicals to ground-water and surface water. Waste products when kept uncycled add foul smells, unsighliness, additional dredging costs of water-ways, silting up of reservoirs, decrease in plant productivity, corrosion of structures and structural foundations and depreciation of land value.

Industries take the major blame for waste piles. A 1000 mega watt thermal power station piles up 3 meter deep ash on 20 hectares area in its life period of 40 years. Metal industries dump 50 percent of the ore volume they process or solid-waste. Consumer with snobbery for elegant packing of utility goods, foolish use of paper from elementary school to governmental correspondence practices, short lived consumers goods of use and throw-type and junk discarded due obsolescence add up to the municipal garbage. Construction debris and frequent demolition of structures to be replaced by more commercial profitable ones also add-up municipal solid wastes.

Individual and Global Environmental Ethics

Sanitation is the most sacred of all the activities performed by man. Aesthetic factors and ethics are the most important aspects in sanitation. Clean community with clean surroundings is a way of life. Physical health and development rest on man's physical environment. This is controlled by the water that man drinks, the food which he eats, the house in which he lives as well as disposal of wastes and other wastes matter in community. As human populations grow economic activities grow simultaneously leading to a number of environmental problems that threaten not only human health and the productivity of ecosystems, but in many cases habitability of the globe. Motor vehicles cause air pollution, petroleum refiners emit gaseous pollutants, fly-ash and hydro-carbons by thermal power stations, inorganic and organic pollutants presented by industries, chemical through agriculture discharges, phenyls, toxins and acids released by dying textiles, paper and pulpmills. Are they not creation of human activities. This show that each us is individually responsible for the quality, for better or worse. There are no two opinions over it that man and man alone is responsible for polluting the atmosphere. This show that each of us is individually responsible for the quality, for better or worse. There are no two opinions over it that man and man alone is responsible for polluting the atmosphere. This does not mean that he should stop performing everyday activities. What needed is real changes in individual behaviour. Environmental ethic can modify his attitudes and behaviour. Environmental ethics should not only be expressed itself in national laws but profoundly changing individuals the way they live their daily lives.

Americans think environmental problems can be solved through a quick technical changes. Raper Organization on environmental problems believe that cars, not drivers pollute environment, hence corporations should invent pollution free autos. This shows that many individuals want the environment cleaned up, but they do not want to make life-style changes to make that happen. Decisions and actions by individuals want the environment cleaned up, but they do not want to make lifestyle changes to make that happen. Decisions and actions by individuals when faced with ethical choices can collectively save the environment from further degradation.

Changes in individual behaviour are necessary but not enough. Increased awareness of global environmental change is important. This can be done by pooling our knowledge and coordinating our actions. Environment is a matter of global priority and action.

Whatever political differences between different nations have must be dropped/ After all, ecological degradation in any nation inevitably impinges on the quality of life in others. Studies reveal that much of the environmental crisis are rooted due to widening gap between rich and poor nations. Developed industrialized countries contain hardly 23 percent of the world's population and control 80 percent of the world's goods.

Environmentalists hold these countries responsible for pollution. Underdeveloped developing economies are hit by overpopulation, malnutrition, and disease. These nations struggle to catch up with developed nations. They make rapid efforts towards industrialization which poison their cities and attempts to increase agricultural production often result in feeling of trees which deplete the fertility soil. So the most important question of the future is put aside political differences and work together for global environmental course of action. The United Nations conference on Human Environment was a step towards this direction in 1972.

It is agreed the interactions between man and environment are not new. They are as old as human civilization. In past such interactions were less but now unprecedented increase in the rate, scale and complexity are observed. At one time pollution of any nature was a local problem but today it involves several countries. The discussions and debates over chemical and radioactive waste disposal are examples of international nature of pollution. The Gulf War which took place between Iraq and Kuwait in 1992 polluted the atmosphere. Pollution from the burning of about 3 million barrels of oil a day produced a pungent lmog. The polluted atmosphere created a very bad impact on human health. Reports reveal that babies suffered from lung problems; adults suffered from high blood pressure, physiological disorders, kidney failures and heart attacks. The burning wells produced huge amounts of smoke turned days into nights. The wind screens of the vehicles were found carrying black sports and in some cases turned grey and became untransparent. There were black rains in several parts of the world. The snow that fell in Kashmir during the post war period had black spots due to hydrocarbon pollutants. Should we not think less about conquering nature and more about learning to work with her.

2

ENVIRONMENT OF COMMERCE

The main function of commerce is to make the goods available to the consumers. Trade will not come into being unless goods are produced in excess of local needs. Thus, development of trade depends upon production of goods and services on a large-scale. The production of finished goods in the factories depends upon production of raw materials products. The primary products of raw materials have no utility or very limited utility in relation to man, for example cotton produced by the farmers to the textile mills. Moreover, cotton sold by the farmers needs to be transported to the textile mills. The finished goods produced by the factories need to be distributed to the different market places through different modes of transport. In the market areas goods are sold by the retailers to the consumers, which they purchase either from wholesalers middle men or directly from the producers. As wholesalers deal with the large-scale purchase and sale of commodities they are often required to use services of banks, warehouses, financial institutions, insurance and government offices. All these activities are necessary for the development of commerce.

Hence, these activities constitute environmental of commerce. Physical and man-made elements around man, constitute his environment. Similarly, the activities and services necessary for the smooth functioning of trade, constitute the environment of commerce. Variation in the development of these activities and services create commercial environment of their own king. Commercial environment of rural areas is different from that of urban areas.

The Service Sector Activities

The service sector activities can also be classified into two categories :

(i) Quaternary activities : These activities provide necessary infrastructure for the development and smooth functioning of all above mentioned activities.

(ii) Tetairy activities : These activities are concern with trade and aids to trade such as retailing, wholesaling, banking, insurance, transportation and communication and communication services etc.

Productive Sector Activities

The activities which are directly concerned with production of goods came under this category. These activities on the basis of the nature of products produced by them can be further classified into two activities, i.e., primary and secondary activities.

(i) Primary activities : These are the activities producing the products having raw or primary nature such as food gathering, hunting, fishing, grazing, farming, lumbering, mining etc. Through these activities man directly gets the raw materials or primary products from the nature.

(ii) Secondary activities : These activities are concerned which conversion of all kinds of primary products into a variety of useful of finished products.

Factors Influencing Environmental of Commerce

We have recognised the different forms and levels

in the trade organisations. The location and development of different trade organisations and the consequent development of commercial environment depends upon variety of factors such as physical and cultural factors. The nature of commercial environment in any region or in any country is greatly influenced by the physical factors such as availability of natural resources and their potentiality, level of development of agriculture mining, fisheries, forest and animal wealth etc. However, the role played by man-made factors in influencing the nature of environment of commerce is of vital importance.

Man-made or cultural factors such as size and quality of population, political and economic systems, government initiative and encouragement, play important role inn deciding the nature of environment of commerce. Let us study how these factors operate and affect the environment of commerce.

1. Man plays a unique role in the development of environment of commerce, being both producer and consumer of various goods. Man is the thicker, the planner, in inventor, the discover, the coordinator manager, the dreamer and above all the superior. He is an inexhaustible source of energy and skill on this planet. With all these tools he has created his own cultural world, a world of commerce and trade by bringing about revolutionary changes in the field of production, distribution and consumption of various goods of commodities.

2. Political Stability and Government Policy - They also influence development of commercial environment in a country as well as in the world. War between two countries Arab- Israel, Iran-Iraq not only affects the trade between such countries world trade in mineral oil was

adversely affected.

Government is powerful institution and plays a vital role in the development of commercial environment in a country. The Government, by its policy, is empowered to change the pattern of economic progress and trade through administrative process. It also controls credit it institutions, licensing import and export and trade process. It also controls credit institutions, licensing import and export and trade process. The form of government gives all kinds of freedom in trade and commerce and thus crates favourable environment for the development of commerce and trade. This is not so in case communist countries like China where trade activities are fully controlled by government.

3. Transportation and Communication Facilities - Development of trade and commerce is promoted by varied and efficient transport network. West European countries and eastern central pair of North America have become the leading commercial regions of the world, largely due to their dense network of roadways, railways, airways and even inland waterways. They also have the world's best communication systems. In contrast to this many African, South American and South Asian countries have remained commercially backward. They failed to create suitable commercial environment due to lack of these facilities.

4. Existence of Innumerable Differences in Human Needs, Wants and Desires - Human wants are ever increasing and never ending. These ever increasing wants of man provide scope to produce goods and services on a large scale. Consumer especially from urban areas in general and from western countries in particular have very high expectations. Technology has want more of many things than more of the same thing.

New varieties of products, superior in quality, free from pollution, more safe and comfortable, are to be produced and supplied to the affluent sections.

This calls for a massive investment in Research and Development. Japan is a fine example in the regard. Japan always takes care its customer's high expectations regarding design, sophistication, quality, delivery schedules and prices. Industry owners in japan swear by the dictum The Customer is God, who is always right. This attitude has enabled Japan to develop conducive commercial environment, making it one of the leading commercial nation in spite of lack of raw materials in the local areas.

5. Availability of Natural Resources and the Level of their Development - Development of trade and commerce, hence the development of commercial environment, depends upon production base of a country i.e., how much is the production of primary and secondary goods in country. Large-scale production of primary and secondary goods in a region of in a country depends upon its richness in commerce and trade, as they are rich in variety of natural resources such as cool energetic climate, fertile soils, basic minerals like coal and iron-ore, marine and animal resources. Moreover these countries being culturally well developed, have utilized their resources to their fullest capacity. In contrast to this, many African and South American countries have remained commercially backward due to unfavouable nature of their natural resources coupled with their cultural backwardness.

Consumer Space

From the above discussion it becomes clear that consumer space represents that section population space, which provides demand for certain goods.

Population space is a wider term. It includes all the people living in a particular region or country. All the people of a country or of a region do not purchase all the goods. They purchase goods as per their purchasing power, needs, liking, habits etc. In case of essential goods slide food grain, edible oil, cloth etc. the entire population space becomes the consumers space.

But this is not so, in case of luxury goods like, refrigerator, television etc. Thus, in case of higher order goods consumer space becomes more specialized and selective. It is because of this, while locating shops dealing in luxury goods, consumers space need to be taken into consideration. It will not be possible and profitable to open television and video shops in the villages or in tribal areas. Thus, while locations a shop or starting a business the study of consumer space becomes necessary.

We have studied that people in advanced countries enjoy higher standard of living. Their purchasing power as well as of consumption are very high and hence, provide demand for a variety of goods. Same is the with the people living urban areas. Hence, shops dealing in luxury goods are found well developed in these areas.

In U.S.A. level of consumption of people is so high that in spite of having large-scale production of milk and milk products, they import butter from Denmark, Netherlands and New Zealand. The temperate countries of southern hemisphere such as New Zealand, Argentina, South Africa and Australia export dairy products to the temperate countries. The people in tropical countries being economically backward cannot provide demand for costly dairy products. Thus, consumer space is very large in developed countries, though, their population space is comparatively smaller.

In contrast to this, in developing countries though population space is larger, consumer space is restricted. Similar condition are also observed in urban and rural areas. Therefore, while selling goods in the marked knowledge of consumers space is necessary. Nowadays consumer is considered as a king of God. His opinions, likes and dislikes, fashion guides etc. direct productions in the industries. If a consumer happens to dislike a particular product, its production, sour or later comes to an end. People in advanced countries as well as in urban areas have a variety in their demands and hence, to satisfy their demands careful study of consumers space is necessary. In fact, success of failure of any commercial organization largely depends upon its better understanding of consumers space.

Population Space

The word space refers to the geographical area. From the commercial point of view distribution of population across the geographical area is of utmost importance. It is because development of different commercial activities area depends upon the level of their demand. Demand for various goods and services depends upon density and distribution of population, i.e., how populations is concentrated and spread over in a particular area or a region. We have already studied, how population is found widely scattered in the rural areas but its is quiet compact in case of urban areas. Similarly, in villages number of people living is far less than those living in cities and towns. It is because of sparse and widely scattered population in rural areas that the organised markets and developed. Thus, a study of distributional and density of population is essentially useful for commerce. It helps us to know where demand for various goods is concentrated and hence, he location of the market. So, the study of he density and distribution of population is very important to understand the commercial environment. However, mere study of distribution, and density of population is not enough because demand for goods and services more precisely depends upon quality of population, i.e., purchasing power, standard off living, liking and disliking, fashions, customs, literacy ratio, male-female ratio, age and occupational structure.

All the above qualities determine the size of market. The size of market does not necessarily mean the number of customers. If that is the case China and India might have become the world's largest markets. Thus, it is the purchasing power of the people which determines the size of the market. For example, the population of U.S.A. Though far less than of India, has constituted one of the largest world. The higher level of consumption in such countries mainly attributes to their higher purchasing power.

In the world there are four major zones of high density. (i) Southern Asia, (ii) Eastern Asia, (iii) Western Europe and (iv) Eastern Central part of North America. In the first two regions, expect Japan, most of the countries are economically backward and hence, people in these countries have poor purchasing power. They cannot provide demand for various goods while in the other two productive capacities, there is always a demand for variety of goods. Thus, size of market does not necessarily depend on size of population but on quality of population.

Location and Development of Trade Organisations

Above, we have seen, how various factors influence, the environment of commerce in the world as a whole. Now we shall consider how different trade organisations get developed on a given geographical space. Also what factors influence their location and development in various areas leading to the development of their own commercial environment.

1. Normally wholesaling activities are found developed in cities and larger towns. It is because wholesalers deal with large scale purchase of different commodities, hence, require large size trading areas. In the cities and in the larger towns population being large in size to cater to the needs and wants of large number of customers, large numbers of retails shops get developed. They provide the necessary large trading areas to the wholesalers and hence wholesaling activities get developed in the urban and semi-urban areas. Moreover innovative urban transport and communication systems help the wholesalers to serve large number of retail customers not only from cities but even from neighbouring areas of cities and towns. Thus ultimately help them to earn large about of profit, thereby providing several facilities to retailers-their-customers-to earn their goodwill who give them business.

Within cities wholesaling activities show area specialization. They are found located close to central business district the most accessible part of a city and not the geometrical centre of the city. It is because, as wholesalers carry on business on a large scale, they are often required to use specialized services like the services to head offices of banks, insurance companies, industries as well as administrative offices etc. These services being meant for the customers, get located at the most accessible part of a city, i.e., C.B.D.

With the passage of time wholesaling activities developed closest to C.B.D. Show real specialization leading to organization of commercial business. We can identify different wholesaling zones in Mumbai such as Phule market for fruits, Kalbadevi road for cloths, Bhuleshwar etc. Here, they are benefited by their agglomeration. Hence, all these services and wholesaling activities form an integral part of the commercial environment which exists at such places in the cities.

Certain towns which are found centrally located with respect to the neighbouring villages producing a particular commodity on large scale as Pachora, Amalner, Dhuel towns in Maharashtra were known for their specification in wholesale trade of cotton. Khanna town in Punjab is known for wheat, Nagpur for oranges and Ratnagiri for mangoes. Thus, we observe areal affinities of certain trade commodities.

(ii) In a metropolitan city like Mumbai, as population is very large in size and having higher purchasing power, many departmental stores and shopping complexes have developed not only in the commercial centre of Mumbai. The areas in Mumbai which enjoy maximum accessibility have attracted location of super markets and consumers co-operative societies.

(iii) The shops dealing in same commodities are often found located at the places which are easily accessible to the large number of customers. Here, customers living on both the suburban railway lines, i.e., Central and Western, can easily come to make their purchases. Such location of retail shops dealing in same product on a particular road is beneficial for both retail shop owners as well as a change to compare quality and make their purchases form such areas. This provides large turnover of business to the retailers. Therefore, the get concentrated on a particular road. This leads to developed of commercial land use which is unique in nature.

Moreover, when customers come to such roads for making their purchases, they often combine purposes of their visits to such areas i.e., customers coming the Ranade read for purchasing clothes. They often purchase vegetables or shoes or ready-made garments and other goods for Dadar, which are very well available in their local areas. It is because of such multi purpose trips many shops dealing with vegetables, ready-made garments, plastic goods etc. have developed on the footpath of Ranade road and adjoining roads at Dadar.

(iv) Retail shops dealing with costly and non-frequently required commodities such as cloth, medicine, shoes, jewellery, radio, T.V. etc. are found located comparatively away form the residential areas. They are mostly found located on the main roads leading to a railway station or bus stand, if any. Once in a while customers don't mind travelling long distances while making purchases of such commodities.

Moreover the actual location of such shops also depends upon rent or cost of the land. The shops capable of paying higher rent or cost of the land are found to be located at the places having two shopping fronts or junctions.

(v) The retail shops dealing in cheap and those commodities required in daily life and found located nearest to the residential areas. This is because of consumer behaviour like customers not ready to travel longer distances for purchasing cheap and frequently required commodities. In fact, if the customers are forced to travel longer distance, to purchase such commodities, the cost of transport may exceed the cost of commodity itself.

Linkages between Producer and Market Environments

Commerce is unthinkable without linkages between producer and consumer. In fact, the linkages between producer and consumer form an integral part of the environment of commerce. These linkages may be in existence in a traditional form, as they where in the past in the form of barter system while transport of goods used to be carried out the help of man himself or animals. However, today these traditional linkages have been replaced by modern linkages such as cash or cheque payments instead or barter system and in case of transport of goods by means of containers, that too, through different modes of transport such as roadways, railways, waterways and airways. Thus, the linkages between the producer and consumer are of different types and vary from region forming their own environment of commerce. For example, in mountainous and desert areas, animals are still used as a beast of burden.

The linkages between producer and consumer are mainly with reference to transportation, payment, distribution agencies and price of commodities. These linkages along with their elements, play an important role in determining the nature of environment of commerce. For example, in advanced countries, banks, modes of transport, sales agencies and price guarantee system are well developed. In developing countries, these linkages are not well developed, especially in rural and tribal areas. This has affected the nature of commercial environment in these countries.

The various components of above-mentioned linkages also play an important role in the development of commerce. The linkages between producer and consumer are not only affected by man-made environment but sometime they also affected by natural calamities, such as heavy rainfall some mountainous areas and snowfall.

Internal Economies and External Economies

Internal economics are those economics which are open to an individual firm when its size expands. They emerge within the firm itself as its scale of production increases. Internal economies in the scale of its output cannot be realized unless the firm increases its output.

External economies are those economies which are shared by all the firms in an industry or in a group of industries when their size expands. They are available to all firms from outside, irrespective of their size and scale of production. They are the result of the growth and expansion of any particular industry of a group of industries as a whole. Thus, external economies are the functions of the size of the industry. They are not confined to one or two firms, but shared by the all the firms in an industry when its size expands. Thus, external economies can never be monopolized by any one in an industry.

3

GLOBAL TRADE

Environment of Commerce

The term trade refers to the exchange of commodities, services and know-how. Trade may be internal or domestic when goods or services are exchanged within the political and geographical boundaries of a nation or region; or it may be international or global when the exchange of goods or services is between two or more countries and across political boundaries. The latter is also known as intra-regional or foreign trade.

The economic progress of any country depends considerably on the development of trade relations with different countries. Different countries produce different types of minerals or crops. It is due to diverse geographical and economic conditions. Countries of the world have come closer due to effective development of transportation. There is no country in the world which is self-sufficient in the production of all foodstuffs and raw materials. Countries only specialize in different products under the particular conditions of physical and economic environment. Trade is the result of specialization which is a function of demand and supply.

Internal and External Trade

The Internal or domestic trade of any country is based on various factors, namely,

(i) Transportation.

(ii) Habits and size of population.

(iii) Differences in climate conditions.

(iv) Differences in the stage of Industrial Development.

(v) Uneven distribution of natural resources.

The fundamental basis of trade at any level is uneven distribution of natural resources. A large size country like India has developed internal distribution systems for coal, oil, iron ore, agricultural commodities and industrial goods. Certain agriculture products are produced in particular areas only, e.g., tea, jute, wheat, rice, sugar, etc. But they are consumed all over the country. There is a good deal of internal trade in these goods. Also we notice that there is an uneven development of industries are distributed all over the country. Sugar is produced more in Uttar Pradesh and Maharashtra. Other states are in deficit and therefore there is good deal of internal trade in sugar. Food habits of people in different parts of a country are different. Rice and coffee and more consumed in Southern India while wheat and tea are more consumed in the rest of India. If these commodities are not produced or are produced in deficit in area then naturally they are brought from the areas of surplus production. Thus the internal trade in these commodities is developed in India.

Size and type of population also helps to promote the internal trade. Larger the size of population of the area, larger the volume of internal trade. The types of the levels of economic standard of people also decide the volume as well as the kinds of commodities traded within the areas. The transport linkages play very important role to promote internal trade. Railways and roadways have been found to be more useful in internal trade of a country. They have not been adequately developed in India.

An external trade or a foreign trade of a country is influenced by following factors:

(i) Political relations.

(ii) A demand for a commodity or product.

(iii) Quality and cost in competitive world market.

(iv) Exportable surplus production of certain product.

The external trade of a country is a part of International trade. This king of trade is developed on account of uneven distribution and development of resources in the world, differences in the stages of industrial development, differences in the culture and skills of people, difference in climatic conditions, habits and tastes of people.

Besides above- mentioned reasons, the external trade is possible if their is an exportable surplus output of a particular commodity in a country and the commodity has a demand elsewhere. The trade develops from areas of surplus to areas of demand. The trade in particular commodity can last in this competitive world if better quality goods are sold at reasonable price. The surplus productions of better quality goods at reasonable price such as coffee by Columbia and Brazil; sugar by Cuba and Brazil; tea by India and Sri Lanka; synthetic fibre electronic goods, automobiles and ships by Japan, therefore, enter into world market on large scale. The demand for a commodity in foreign markets increases the trade of a country. The political relationship plays important role in external trade of a country. The strained political relationship between India and China could not allow to develop any trade during last four decades.

The positive balance of a foreign trade of a country helps to make the economic conditions sound.

Obviously negative balance of a foreign trade makes the economy weaker. The comparative territorial advantage is enjoyed in production of certain commodities by many countries in the world. It is possible for them to produce those commodities at least on cost production basis and in much exportable surplus output. Therefore some countries are specialised in the production and their export trade consists of these commodities, e.g., rubber and tin by Malaysia and Indonesia, tea by India and Sri Lanka, jute any India and Bangladesh, coffee by Brazil and Columbia, oil by the Middle East, sugar by Cuba and Brazil, paper and wood-pulp by Canada, etc.

World's Trading Pacts

The term trading pact or trade blocks refer to a group of two or more countries getting together for the purpose of promoting international trade. These trading blocs help to liberalize trade by lowering of abolishing tariffs. Economic integration initially helps to develop political integration. There are many advantages of forming trade blocks such as:

(i) The formation of a group for trade purpose has led to political unity. Apart from economic and commercial safety, political co-ordination and safety are two of the achievements of a trade bloc.

(ii) Free trade policy facilitates and encourages greater specialisation in particular field on production. A trading bloc may become specialised in agricultural or industrial goods. Sometimes it enjoys the territorial advantage because of its geographical location or geological structure.

(iii) Since trading restrictions are minimised, internal trade among the member countries and, therefore, each bloc is regarded as a single unit. Trading blocs and individual member countries, resulting in increased volume of trade.

(iv) Tariff concessions and abolition of trade restrictions help to promote trade among member countries. The volume of trade within the group has been growing very rapidly.

There are many trade blocs in the world. Some of the important ones are as under:

a. United Nations Conference on Trade and Development - It is the biggest pact ever entered into for trade and development at the international level. The organisations was set up to find out ways and means of expanding trade and upgrade the socio-economic status of people in the underdeveloped areas of the world. In 1964, when it was set up, UNCTAD has 120 members. Now there are more than 130 members. The problems of international trade and economic developing countries. After several conferences, it was decided to develop preferential arrangements of tariff in advanced countries to promote the export of several commodities to them by less developed countries. UNCTAD, therefore promotes the trade and development of all the developing nations of the world.

b. General Agreement of Tariffs and Trade - A pact was enter into in 1946 under the patronage of the U.N. to frame a charter for the International Trade Organisation in London. Seventeen countries including India were represented. The pact, however, was given shape at a meeting of 23 countries participating at Geneva in 1947. The GATT not has 86 members and its aim is to reduce tariffs and encourage world-wide trade links for mutual benefits. The headquarters of GATT is at Geneva in Switzerland.

c. South Asian Association for Regional Cooperation - It includes the countries of South-Asia such as India,

Bhutan, Pakistan, Nepal, Bangladesh, Maldives and Sri Lanka. One of these countries is being chosen as the chairman for a period of one year and every month an audio-visual film is shown on T.V. on the 1st day of every month. It is for promotion of cultural and commercial relationship among all the countries of the region.

d. Association of South-East Asian Nations - It is an association of South-East Asian nations, namely, the Philippines, Thailand, Malaysia, Indonesia and Singapore. Tariffs between these countries have been reduced, therefore, intra regional trade has increased. Simultaneously their external trade has increased very rapidly. Except Singapore, all the member countries are producers of agricultural products and industrial raw material. They supply these products to Japan, European countries, Australia and New Zealand. All of them import manufactured goods from Japan, Australia, Korea and European countries.

e. Latin American Free Trade Association - This organisation was formed in 1960 under the Treaty of Montevideo. All the South American Countries, excepts Surinam and Guineas, are members of this pact. Mexico is also a member. These is not much trade among the member countries because of the similarity of their products. It is not more than 10 percent of the total trade of Latin America within LAFTA. Most of the countries have trade links with the industrial nations in the northern hemisphere. LAFTA has either lowered' or abolished tariffs on agricultural products and assisted in the economic development of the sub-continent. A sub group of LAFTA from the Andean region Chile, Peru, Bolivia, Columbia and Ecuador has registered success in trade and economic integration and it is known as Andean group.

f. European Coal and Steel Community Bloc - A trading pact entered into by Belgium, Netherlands, Luxembourg, West Germany, France and Italy, to promote iron and steel industries in member countries by abolishing tariffs on iron and coal. The policy might help in developing monopoly in the production of iron and steel and the supply of steel in world markets. As a result of this pacts, there has been remarkable development in their iron and steel industry because of the exchange of material and technological know- how.

g. The Organisation of Petroleum Exporting Countries - The organisation cane into being in 1960. Thirteen major oil producing countries except, U.S.A. And the former Soviet Union, became its members - Saudi Arabia, Iran, Iraq, Nigeria, Gabon, Equator. Many of these countries have small-size populations, but are very rich in oil reserves and produce large quantities for exports. The organisation takes care of member countries and assists them in getting a better price for their oil by increasing or decreasing production, fixing the price for their oil by increasing or decreasing production, fixing the price of oil, and by giving advice on economic and technical matters. OPEC has shaken the economies of world's leading nations by increasing the price of oil in the world market.

h. Council of Mutual Economic Assistance - A trading partnership of former Soviet Union, Mongolia, Former East Germany, Poland, Czechoslovakia, Hungary, Rumania and Bulgaria, a group of communist countries, was establish in 1949. The former Soviet Union has greatly benefited by their partnership because she was the most advances country in manufacturing activities. She exported coal, oil, machinery, automobiles, mining equipment, agricultural machinery, transport equipment and chemicals to East European countries. She imported fruits, vegetables, textiles, etc.

(i) European Free Trade Association - This is an association of eight European countries: Iceland, Norway, Sweden, Finland, Switzerland, Austria, Protagon and the Faroe Islands. Iceland joined the pact in 1969, years after its formation. The United Kingdom and Denmark were founder members of the EFTA but left it to join the EEC.

In terms of the EFTA member countries agreement, tariffs on manufactured goods have been substantially reduced or abolished. This has promoted the industrial development of member nations and attracted foreign investors- especially American- for the development of industries in these countries. Its internal trade, however, did not develop substantially, may be because of the similarity of their products. The member countries of EFTA have strong trade links with the EEC and other countries.

j. The European Economic Community - This is an association of countries which has progressed very much in trade and general economic integration. It was formed in 1957 under the Treaty of Rome. There were only six members in the beginning- France, Germany, Italy, Belgium, Netherlands and Luxembourg. The trading conditions were changed in 1973. At that time, Britain, Denmark and Ireland joined the EEC. Norway became a member in 1981. Greece, Spain, Portugal and Turkey followed suit.

Though established in 1957, concessions in tariff and trading restrictions were not abolished immediately. But most restrictions on trade in industrial goods were abolished in 1967. However a common external tariff has now come into operation. The overproduction of agricultural and animal products because of the policy of interrogation of agriculture development by the EEC sometimes create problems. But the EEC has been successful in overcoming these problems. Therefore, it accounts for 33 percent of the world's trade and about 45 percent of the trade in manufactured goods. Five members countries are, of were, colonial powers, which helped in promoting the world-wide trade links of the EEC.

Basis of World Trade Patterns

(i) Political Influence on Trade - There has been a great political influence on world trade. Healthy political relationships between any two countries promote strong trade relationship. The political environments in different parts of the world have resulted into the regional trade carried through the trade blocks or trade organisations which enabled the member trade organisations which enabled the member countries to adopt Free Trade Policy or protected trade policy. Today we notice that the trades flows have emerged from the combination of Geographical, Economic, Social, Cultural and Political Environments at all levels.

(ii) Economic and sociocultural base - The countries with high economic and socio-cultural status are engaged in production of the goods of high production cost. Japan, France, U.K., U.S.A., with modern scientific and technological know how, sound economic base and high socio-cultural background produce exportable surplus sophisticate and fabricated goods as well as export scientific and technological know-how to rest of the world. Japan is specialised in electronic goods, automobile, chemical, shipbuilding, synthetic textiles etc. European countries and U.S.A. Are also specialised in automobiles, aircraft, chemical industrial plants and capital goods etc. A large surplus is produced in Asia, South America and Africa are the ready markets and very well linked by international trade routes. The high cost consumer goods produced by Japan, European countries and U.S.A., have good demand in the oil rich Middle East while other capital goods, fabricated countries. The multinational companies are also reaping heavy profit from developing countries- because of their dense population and high consuming capacities.

(iii) Geographical Base - The various commodities that enter into the world trade have geographical or natural bases. Among the food products, foodgrains and animal products, fish and fish products are produced in exportable surplus quantities for trade by some countries. Food grains were important commodities in the world trade till the second world war. The world's food production has outpaced population growth. During last decade the world's food production increased by 24 percent while its population grew by 20 percent. It is noticed only in Asia, Europe, South America and C.I.S. The food production increased because of increase in irrigation, mechanization, inputs of food grains and other products in developing countries are left for export because of large size populations. Rice producing nations of Asia are leading consumers of rice so, there is little trade in rice. Trade in wheat has also declined because developing countries have reacher a stage of self- sufficiency. Only developed countries, export wheat and maize to Japan, European countries and other deficit producers. Where is a major food item enters world trade. The U.S.A., Canada. C.I.S., France etc.

The industrial raw materials include agricultural products like cotton, jute, tobacco, minerals in the form of ores of semifinished products. Most of these industrial raw materials are produced in tropical countries and had a great demand after industrial revolution in the U.S.A. And Europe. The demand did decline after second world war. The new inventions by Americans and Europeans enabled them to produce substitutes. Moreover many tropical countries and developed countries has been declining. The industries in developed countries depend on others, especially, underdeveloped countries suffer for short supply of raw materials.

Broad Geographical Flow Patterns

Almost all countries are produces of primary goods which include minerals, agriculture products like food and industrial raw materials. The developing countries among tropical group have borrowed of hired or purchased scientific and technological know-how and started processing the natural resources they possess. Therefore either semiprocess or processed products are exported by them. Some of the developing countries like India export tools, machineries and engineering goods. She is one of he ten most industrialize nations in the world. However, most of the tropical countries export low value primary goods like agricultural commodities and minerals to temperate countries.

The temperate countries cannot produce certain agricultural products which need tropical climate as well as they do not possess industrially useful minerals. The import them but being industrially advance countries, iron and steel goods, transport equipment, machines and tools, chemicals, etc. The kind of flow of raw material and unfinished products from tropical countries to temperate countries and a flow of manufactured goods or finished products from temperate countries to tropical countries will continue till there is a gap in industrial development.

Northern Hemisphere and Southern Hemisphere

A flow pattern of trade between the countries of northern hemisphere and southern hemisphere is of a typical type. The countries of southern hemisphere are thinly populated and have poor industrial development. They are producers of agricultural commodities, animal products and minerals in exportable surplus quantities. Most of them were European colonies. They are all exporters of primary and unfinished goods to the countries of northern hemisphere. Foodgrains by Argentina, Australia; wool, meat, diary products by Australia, New Zealand and Argentina; Sugar, coffee by Brazil; coffee by Columbia; iron-ore, manganese, copper by Australia, Brazil South Africa; bauxite by Australia, Brazil; tin by Indonesia, Bolivia. The countries, e.g., Japan, European countries, U.S.A export machinery, industrial plants, chemicals, iron and steel goods, automobiles, aircraft, petroleum, petroproducts, textiles and variety of manufactured goods. Some of the countries, e.g., Brazil, Argentina, Australia are now developing modern industries to reduce the import of manufactured goods.

Trade Flow in the Last Two Decades

There has been a major change in pattern of trade flow between developed and developing countries during last two decades. There is a vast difference in the industrial development between these two groups in the world. The developed countries have achieved a very high stage of industrial growth. The developing countries in the world are generally termed as third world countries, mostly the tropical countries in Asia and almost all countries in Africa and South America.

They are industrially less developed. The export trade of developed countries consists of manufactured and fabricated goods the developing countries consists of agricultural commodities, industrial raw materials, including minerals, a relatively large share of their GNP. This trade has a lower economic and political countries of the world have dominated world trade, accounting for about 66 percent while developing countries contribute about 24 percent. Some of the developing countries have developed industries, therefore the trade pattern of those countries have changed during last 20 years i.e., India, China.

Euro-American and Asia Countries

The Ero-American countries are highly industrialized in the world. U.S.A. And Canada are rich in resources. U.K., France has colonies in Asian countries with a long time commercial ties. There is a flow of manufactured goods like chemicals, iron and steel goods, industrial plants, automobiles, petroleum and petroproducts, newspring, paper-pulp, variety of manufactured goods and scientific and technological know-how from Euro-American countries. On the other hand a flow of unfinished and semi-finished goods from Asian countries, except Japan is continued fro last more than two centuries. However, an import of finished goods by India and China has declined on account of progress in manufacturing after post-war period. Petroleum and petro-products from south West Asian countries are exported to Euro-American countries on large scale.

The Flow India to S.E. Asia, Africa and I.O. Islands

There has been good deal of trade between India and countries of S.E. Asia, Africa and Indian Ocean Islands. There have been separate agreements with these countries. Indo-Burma agreement, indo-ori Lanka

Trade Agreement the Indo-maldives Agreement. The major export by India to S.E. Asian countries and I.O. Islands includes of wheat, iron and steel goods, textiles, hosieries, transport equipments, electrical goods, machines and tools, etc. India imports from them rubber, copra, coconut oils, graphite, dry fish, palm oil, etc. India's trade with African countries is limited, but measures have been initiated to explore possibilities for increasing trade with these countries. Already joint agreement, bilaterial co-operation , joint ventures have been signed to establish trade link with them. There are offices of India trade agencies such as the Engineering Export Promotion Council at Abbidjan, the Trade Development Authority in Liberia, the Projects Equipment Corporation in Harare in Zimbabve and Ethiopia. There is ample scope for the export in manufactured goods to these countries and activation balance of trade with them.

Foreign Trade of India

Brief Review

In the pre-independence period, the pattern of India's foreign trade was traditional- the import of finished goods and the export of agricultural produce and industrial raw material. The bulk of her foreign trade was with Great Britain and Commonwealth countries. In view of the size of the country, and the size of her population, the volume of India's trade is very low. The reason are:

(i) The large size of the population, which used up some primary products, leaving a small surplus for export.

(ii) The near absence of roadways and railways connecting hinterland with ports to promote the export of unfinished products and the import of finished products.

(iii) Colonial period of about three hundred years, during which India was compelled to be a producer of low-priced primary goods for the colonisers.

(iv) ignorance, illiteracy and backwardness in scientific and technological know-how; hence, the absence of a scientific approach.

(v) The slow rate of development in agriculture and industry, resulting in low production.

In the post-independence period particularly from 1951 onward, there has been a remarkable image in the pattern of the foreign trade of India. She has now established trade links with practically every country in the world. She has now a fairly large volume of foreign trade and ranks sixth among the trading nations of the world. Her import of manufactured goods has been steadily declining; but her exports are on the increase. This change has come about because of improvement in manufacturing techniques and encouragement from the government. Her trade with U.K. has declined while that with U.S.A., Far Eastern countries, and the Middle East, in on the rise.

North/South

The pattern of trade flow between the countries of north hemisphere and south hemisphere more of less similar to that of developed and developing countries of the world, with few exceptions. The countries of south hemisphere are those in South America, half of Africa, Singapore, Indonesia, number of islands in South Pacific and Oceania. Except Singapore and Oceania all are economically and industrially backward countries. Their export trade consists of agricultural commodities, minerals and animal products. Similarly, their trading patterns in north hemisphere are mostly Japan, Russia, European countries, Canada, U.S.A., which export all kinds of manufactured and fabricated goods to them. The flow of animal products from the countries of southern hemisphere to the countries of northern hemisphere is remarkable.

South/South

There have been limited trade linkages among the countries of southern hemisphere. Most of them produce similar kinds of commodities. Many of them are thinly populated and maintain low standard of living, except Singapore, Australia and New Zealand. They have nothing special to exchange with each other. They are widely separated from each other by waterbodies, noon falls on he line of world trade route.

Pattern of India's Exports

In the pre-independence period, the exports of India consisted of primary products- generally agricultural products, such as tea, jute, cotton, species, tobacco and agro-based industrial goods like textiles, hides and skins and mineral ores. Some of these goods still enjoy prime importance in India's export trade trail development in the fifties, the pattern of India's foreign trade has changed. The number of commodities exported by her at the time of independence was only fifty. Now it is more than 3,000 and includes traditional goods like jute, tea, cotton, cotton goods, tobacco, hides and skins, lac, mica iron ores, magnese ores and spices, and many semi-processed and finished goods, engineering goods, transport equipment, etc. The sixties witnessed the development of industries which manufacture engineering goods, automobiles, other transport equipment, machines, tools, electorial goods and machinery, industrial plants for sugar, chemical industry, pharmaceutical industry, petrochemical industry, cement industries, paper industries, textile machinery and power plants. The exports have not yet wiped out our trade deficit. India is now among the first ten nations in the export of machines and tools. During the last twenty years there has been an enormous increase in the export of engineering goods, leather goods, chemical, jewelleries, readymade garments, handicraft goods, fish and fish products, sugar, and cashewnuts.

Until independence India's foreign trade balance was favourable. However after the independence, the trade balance has been unfavourable, except in 1972-73 and 1972-73.

Essential Features of India's Foreign Trade

The foreign of India is mostly seaborne, and takes place from major ports on the eastern and western coasts. There is very little trade by land routes because of the barrier of the Himalayan ranges. Moreover, the political relations with neighbouring countries are not very friendly.

The per capita share of India's foreign trade is much lower than that of other countries India has hardly one percent the world trade while the U.S.A. Has 16 percent, United Germany 10 percent, Japan 8 percent. Since independence, the foreign trade of India has been growing. The total value of her foreign trade in 1980-81 was Rs. 21, 413 crores and in 1989-90 was Rs. 63,093 crores.

Import Trade of India

In the pre-independence period, the import trade of India covered foodgrains, manufactured goods, machines, tools, electrical goods and transport equipments- foodgrain imports because of the large population and insufficient output, while that of other goods because of poor industrial development.

Greater emphasis was placed on improvement in agriculture by introducing land reforms and irrigation facilities, and popularizing the use of fertilizer and chemicals, high yielding varieties of seeds, mechanizations, and modern methods of farming. By the seventies India became self-sufficient in foodgrain. The imports of foodgrain declined. The process of industrialization has stated by that time. Initially, the machinery for the manufacture of agricultural equipment, machines and tools, transport equipment and industrial plants and their accessories were imported. Then, scientific and technological know-how to manufacture machines, tools and equipments. But industrial plants for heavy industries, such as the iron and steel industry, the petro-chemical industry, the automobile industry, were imported even in 1987-88. India has acquired knowledge in certain areas and will become self-sufficient soon. Industrial plants for the sugar industry, the cement industry, the paper industry, etc., are now made in India and installed by Indian technocrats.

The imports of petroleum and petro-products account abut 50 percent of India's import trade. The rise in the cost of these commodities in the international market has placed a heavy burden on India's economy. Therefore, she was forced to embark on oil exploration. Though she has not become self-sufficient, the pressure on the economy exerted by the import of oil has declined to some extent. In the eighties, 75 percent of her requirements of oil has been met from indigenous sources. But there has been an increasing demand by the increasing population. The imports of iron and steel, metals, certain chemicals, machinery and sometimes of essential commodities like foodgrains and edible oils still continue because of the increasing in the population.

New Trade Policy

The reversal of trends in exports growth during the last two years forced the Indian Government to announce the new export-import policy. The country-faced a serious economic crises with adverse balance of payment position and high inflationary pressure on the price level. There had been a steep decline in exchange reserves and there was not enough foreign exchange to earn import essential commodities and to pay off external debt. To bring the country out of such a difficult economic situation, the rupee was depreciated and it was followed up by major reforms in Trade Policy to boost exports.

In a world of growing interdependence and rapid technological advance, import substitution could no longer be followed at any cost. India has to become more modern, more efficient and more internationally competitive. Global developments like the oil crises of the early seventies and the early eighties forced India to adopt a pattern of development which would enable her to enter the export markets aggressively and earn foreign exchange needed to finance her essential imports. Last year, imports had gone up by 21-6 percent as against the growth in exports by 17-9 percent and the trade deficit reached a record level of Rs. 10,642.54 crores and hence, it was thought that unnecessary controls affecting the country's efforts in becoming globally competitive economy, should be removed. As a result of this the new export-import policy was announced by the Government in Parliament on 31st March 1992. The following are the highlights of the new export-import policy.

(i) The Ministry of Commerce has set up a High Level Committee on Trade Policy Reforms in order to progressively eliminate licensing and quantitative restrictions on capital goods and raw materials, components so that all these items can be placed under OGL.

(ii) Rationalization and simplification of the tariff structure has been undertaken in order to more towards international competitiveness.

(iii) EOU and EPZ schemes have been liberalised in order to attract foreign investment for tapping export markets.

(iv) The new policy aims at progressively reducing the extent of canalization. The Government has decontrolled 116 items, allowing their exports without any licensing formalities.

(v) The new trade policy includes measures to strengthen the development of export houses and Trading Houses in order to promote exports. The range of items permitted to be imported by Export Houses, Trading Houses and Star Trading Houses has been widened.

(vi) The procedure for import of capital goods has been simplified. New units and units undergoing substantial expansions will be automatically granted licences for import of capital goods. The negative list of imports is the smallest-ever. Consumer goods will continue to be under restraint.

(vii) The system of advance licences has been introduced to provide the exporters with duty free access to the inputs they need to produce competitively for world market.

(viii) The system of supplementary licences has been abolished.

(ix) The new trade policy makes major changes in the import licensing system by replacing a large part of administered licensing of imports by import entitlements linked to export earnings. The import replenishment system was enlarged and restructured and renamed as Eximcrips. Under this system all the export products of export services has seen enlarged. Eximcrips will be tradeable and can be used to import any item in the negative list of imports.

Indo-U.S.A. Trade -The large volume of India's trade is with U.S.A. In 1990-91 India's exports of the U.S.A. were 14-7 percent, while her imports in 1990-91 were of the order of 10-9 percent.

The major items of India's export to the U.S.A. are readymade garments, cashewnuts, diamonds, leather goods, jute goods, carpets, drugs, chemicals and allied products. She has started exporting many non-traditional items, including manufactured goods. Her major items of import from the U.S.A. are petro-products, chemicals, fertilizers, iron and steel, machinery, capital goods, spares, components and scientific and technological know-how. Her balance of trade with the U.S.A. is now been favourable.

Canada is another North American country offering a good market for the import of non-traditional Indian goods, The major items of her exports to Canada are textiles, engineering goods, chemicals, leather goods, tea, spices, jewellery, coffee, handicraft, etc. Traditional raw material is being increasingly replaced by finished goods. The imports to India from Canada consists of paper, newsprint, metals, etc.

Trade with South American Countries

India's foreign trade with Latin American and the Caribbean countries is hardly 6 percent of her total world trade. One of he reasons for this low volume in the distance. Nevertheless, India has forged trade links with Mexico, Venezuela, Columbia, Trinidad, Tobago, Cuba, Brazil, Argentina, Peru, Chile bad Guyana and the trade with them is increasing comparatively in the recent years. The major items of her exports to these countries are tea, jute goods, textile goods, machinery etc. India has no trade treaty with any individual country situation. There is a need for publicity for India's products and export capability inn the South American countries.

Trade with African Countries

India's trade with African countries is limited. But measures have been initiated to explore possibilities for increasing trade with these countries. Already, joint ventures, bilateral co-operation and agreements have been signed to establish trade links with them. There are offices of Indian trade agencies, such as the Engineering Export Promotion Council at Abiding, the Trade Development Authority in Liberia, the Project Equipment Corporation in Harare in Zimbabwe and Ethiopia. There is ample scope for the export of manufactured goods to these countries and for achieving a favourable balance of trade with them.

Trade with East-Asian - and Oceania

Among 27 countries in East Asia and Oceannia, only a few have trade-links with India- Japan, Korea, Australia, and New Zealand. Till 1978-79, the balance of trade with this region was favourable; thereafter it has become unfavourable mainly because of bulk imports of palm oil, woo, electronic goods, Natural Rubber, Zinc, Lead, Fertilizer, Coking coal etc.. Japan, of all the countries in trade with Japan recorded an all time high figures of Rs. 5547 crores manganese ore, mica, tobacco, leather goods, etc. Her imports from Japan cover silk, artificial silk, chemicals, electrical and electronic goods, machinery, iron and steel goods etc. There is ample scope for the export of Indian engineering goods and manufactured products of East Asian countries and Oceania.

Joint ventures and trade agreements with other countries have been functioning very efficiently and promoting India's foreign trade as well as increasing the country's participation in international trade.

4

FOREST RESOURCES

Natural Resources

Natural Resources means resources established by nature only. They are natural endowments in the form of land, water, vegetation and minerals. All natural materials with the passage of time become natural resources which provide base for economic development. By utilizing, natural resources, humans created their own world of living. They created houses, buildings, roads, railways, villages, towns, cities, machines industries and several other things. These are also very useful, and hence are called human made resources. Both natural and human-made resources are essential for living.

Resources have certain important characteristics as stated below:

(i) The help us to create goods and provide services.

(ii) The posses utility.

(iii) The are normally available in limited quantity.

(iv) Man has to exercise efforts in order to get utility from resources.

(v) Utility of resources change with the improvement in science and technology.

FOREST

Thirty percent of land area in the world is forest land. Trees are the most dominant living organisms in forests.

Forests function as huge biological filter systems that clean the environment by removing impurities from air and water. They restore oxygen in our atmosphere through photosynthesis, besides providing solvents, medicines, fuels and many other products that are important for our health and comfort. Briefly, forests play a vital role in regulating climate, controlling water run off, providing shelter and food for wild life. A part from it, forests have scenic, culture, and historic values that deserve to be protected. Forests have great biological and commercial values. These values of forests are discussed below:

Biological Values - A forest consists of an area where trees are the most dominant living organisms. The majority of forests are made up of different king of trees. Several layers of vegetation called strata are found in forest. These stratus are canopy, understory, shrubs, shrub layer, herb layer and forest floor. The tall broad leaf trees from the canopy are called ceiling. The areas beneath it is called understory. The lower strata of a forest having small woody plants is called shrubs. The shortest plants such as ferns, grassed and flowering plants are collectively called herb layer. The forest floor is composed of a layer of decaying plant materials.

Besides plants and trees that are found in the forests insects, birds, reptiles and fish are also included inn it. So all of the living organisms that found inn forest environments contribute to the value of forest. This is called biological value of a forest. Forests have some influence on the climate in the local area. They provide shade and cooler temperatures inside forest environment during summer months than in surrounding areas. Forest environments also contribute to comparatively warmer temperatures within the confines of the forest. Dense vegetation in the forest restrict the movement of air. This reduces the chill as well as limit evaporation of moisture from the forest floor. In contrast, large amounts of water are lost from plant leaves to the surroundings area. This results increased humidity in the air. This loss of water from leaf surface is called transpiration.

Trees are important to fish habitat in streams. They provide places for fish to hide and trees contribute to cool water temperatures by shading the water from direct sunlight. Trees protects streams from the heat of the sun that flow through forests. This is important to the survival of some kinds of fish such as trout that require cool water temperatures.

Forests provide food for many species of wild animals besides place of shelter. Animals and birds make their homes in forest environments. These predators use materials found in forests to provide dens and nesting places. Plant materials found in the forests is a big source of medicines that are of great use in our day to day life. Attempts are being made by botanists in the forests to identify unexplored herbs. So forests offer the potential for new medical cures. Research is being made continuously to find many new plant materials inn our forests that are valuable to a society.

Commercial Value of Forests

Forest products are important natural resource. A forest is renewable resources that replaces itself when conditions are favourable to new growth. Nearly 42% of the timber harvested in the United States is used for lumber. Lumber is an important forest products due to its widespread use in the construction of homes and commercial buildings. 20% of the wood harvests is still used for fuel in United States.

Wood is the most important energy source in many countries of the world. Energy is released form wood in the form of heat when commission of burning occurs. A recent development, in wood stoves is the use of wood pellets for fuel. Wood pellets are made from wood scraps, sawdust, shavings, and other waste materials. Charcoal and some kinds of coke are special fuels that are produced from wood for industrial and home uses. Biomass is defined as the weight of all the materials inn a tree that can be used as source of energy. Biomass production is measured as net weight of dry matter. Intensive culture of fast-growing trees has the capability of producing high tonnages of biomass when trees are planted in dense stands and harvested after twenty and thirty years of carefully managed growth.

Forests are the most important source of biomass. They constitute 42% of the total available biomass of the earth. Biomass includes vegetation and wastes that contain significant amounts of vegetable matter. Biomass is a modern wood product consisting of wood chips that are dried and burned as a source of e energy to generate electricity. Perhaps the best known use of the forest is the production of timber products.

Forests continue to hold appeal for people who enjoy outdoor environments, and many people use forests for recreation purposes. Some people enjoy the forest as a wild place where wild animals can be observed in their natural settings. Lakes and streams provide opportunities for fishing, boating and other water sports. Outdoor recreation has become a huge industry, and the citizens of many communities located in or near forest lands derive much of their income from sales of food or supplies related to recreational activities. The forest industry in north America has evolved from the harvest and sale of most timbers to more that $ 100 billion each year. Less developed countries, mainly in the tropics, produce more than half of industrial wood but use only 20 percent. Japan is by far the world's largest importer of wood, purchasing about 43 million cubic meters per year or 87 percent of net Asian imports, The average amount of wood used for cooking and heating in sixty-three less developed countries is about 1 person per year, roughly equal to the amount of wood that each American consumes as paper products alone. Many people in poor countries cook over the open fires that deliver only about one-tenth of the available heat to cooking pots. Jiko stove and clay pot used as an open fire. These stoves save forests, reduce air pollution, and save up to 20 percent of household income.

Types of Resources

Resources are classified in different ways namely as natural and human made, renewable and non-renewable, community and national resources. Natural resources are available with nature. Land, Water, Minerals and forests are some of the natural resources. They are of two types: biotic and abiotic: Land water and soil are abiotic whereas forests and animals are biotic resources.

Resources created by humans are called human made resources Engineering, technology, machines, buildings, paintings and social institutions are some of the human resources. Human resources are essential for development of natural resources.

As already stated, resources are further classified into renewable and non-renewable resources. Some resources have ability to renew themselves in a given period of time. Forests and wildlife are renewable resources. Similarity water, solar and tidal energy are renewable resources. Minerals are non-renewable resources. See below the classification Renewable and non-renewable resources:

Non-renewable Sources

Coal, Space of process heat, Natural Gas, Gasoline and fuel oil , Coal gasification, Oil shale, Nuclear.

Renewable Sources

Hydroglectic, Biomass methane, Wind, Solar electric. Attention mush be paid to the following important points with regard to natural resources.

(i) Resources Development - There are some resources, which cannot be used directly to satisfy human needs, One has to make an effort and develop it for human satisfaction. Land has to be cleared of unwanted materials and ploughed in order to grow crops. Water has to be e taken to fields to irrigate lands. Minerals have to be taken out from the earth and smelt before they are used for making machinery and implements. Resource development, thus, is a necessary exercise for its utilization.

(ii) Resources Planning - As the resources are limited and unevenly distributed over the country, their planning is essential. Resources planning is a technique or skill or proper utilization or resources. It comprises,

a. Planning for exploitation or resources.

b. Evaluation in terms of availability of resources.

c. Preparation of inventory of resources.

(iii) Conservation of Resources - Conservation means judicious and planned use of natural resources.

In other words conservation means preserving and maintenance of the environment to meet human needs for production while ensuring that proper consideration is also given to aesthentics and recreation. The reasonable use of the earth's natural resources- water, soil, wild life, materials is a major goal of conservation. Natural resources are a vital part of sustaining human life, and conservation measures are designed to control, manage and preserve them so that they can be used and appreciated to the maximum. The conservation of natural resources include (i) Mineral Conservation, (ii) Wild-life Conservation, (iii) Forest Conservation, (iv) Water Conservation.

Above described resources form a part of our environment. Proper utilization of these may sustain the environment. Inappropriate and reckless utilization and non-renewable resources, let us discuss forest resources which is renewable resources and contribute substantially to our economic development. They play a major role in enhancing the quality of environment. They perform various types of functions. They modify local climate, control soil erosion regulate stream flow, support a variety of industries, provided livelihood for many communities and offer opportunities for recreation. They reduce wind force and influence air temperature, The add to the forest floor large quantities of loaves, twigs and branches, which after decomposition, forms humus. They provide industrial wood, timber, fuel wood, fodder and several other minor products of great economic value. The national forest policy emphasizes the role of forests in maintaining the life support system. Detailed review of forest wealth is described below to know the value of this renewable source of environment, as well types of forest.

Types of Forest

Forest are classified in three types broadly for the

purpose of administration. They are:

(i) Reserved forest, (ii) Protected forests, (iii) Unclassed forest.

(i) Reserved forests are those which are permanently earmarked either to production of timber or other forest produce and in which right of grazing and cultivation is seldom allowed.

(ii) In protected forest, these rights are allowed subjects to a few minor restriction,

(iii) Unclassified forests consist largely of inaccessible forest of unoccupied wastes. Forests are also classified according to density. Namely: dense, open and mangrove forest/

Forest Management

The forest land constitute a massive renewable natural resources that should be available both for present and future generation of citizens. Forest management refers to the planning of sustainable harvests with particular attention paid to forest regeneration. The role of government in conservation and management of forest resources is to manage public forest lands to sustain the yields of forest products in a manner that is consistent with science validated Silviculture practices. Silviculture refers to growth of more and more trees. It is believed by the forest managements agencies in USA that forests are widely used and will always be available to the citizens who depend upon them Studies reveal and records speak that forests are shrinking rapidly. The man causes of diminishing forests are discussed below:

(i) Most countries replant less number of forest: to be converted uses. Due to divergent definitions of deforestation, rates of forest rates of forest destruction exist in many places. Some scientists and politicians while defining deforestation comment that it means a complete change from forest to agriculture and urban areas. Others define that if an area is logged and selective cutting of forests is exercised then natural growth will be rapid. Countries also have economic and political reasonable to hide or exaggerate the extent of their activities. Consequently, more than 20 million ha per year.

(ii) Incendiary fires account for over 1/4 of all of the forest in the World each year. This fire is stated deliberately by people and is called wilful setting fires. The practice accounts for nearly 40% of all forest fires in the region of United States. The objective of this fire is to control pests and to improve grazing for livestock.

Intense fires generate tremendous amounts of heat and by a fire of this type the combine and phloem tissue in the stem and branches and damaged restricting the flow of nutrients and water enforced. Wild fires often inflict serious damage to soils. This is because organic matter is destroyed, causing the soil structure to break-down.

(iii) Loss of control on a prescribed burn spread wild fire in the forest. A prescribed burn is a fire that is started in an established forest for the purpose of removing vegetation and fuel from the forests floor.

(iv) Human activities are equally responsible for most forest fires. A leading cause of these fires is failure to put out camp-fires, cigarettes, and matches before leaving or discarding them in forest.

(v) Volcano activity which is a rare cause of natural fires in the forest cannot be denied too. Volcanic explosion in 1980 demonstrated the intensity of this natural force. The trees within a radius of several miles of the Volcano were all blown down.

(vi) Fires, insects and diseases damage up to one quarter of the annual growth in temperate forests. The main causes of forest fires are lighting, human activities such as cigarettes, campfires, fireworks and so on. Fire types are determined largely by the kinds of fuels that are present. These fuels are of two types Surface fuels and ground fuels. Surface fuels are one located on the surface of the soil. They are dry leaves and plant materials. Ground fuels include combustible materials located beneath the surface. Duff is a ground fuel. A surface fire burns surface fuels consisting of the dry layer of twings, dead branches, grass and leaves that lie on the surface of the ground.

A ground fire burns ground fuels that are found on or beneath the surface of the ground. During dry periods, these fuels will burn, and a fire in a deposit of duff can smoulder for many weeks.

(vii) Another important reason for diminishing forests is huge profit earned by absentee landlord's and government agencies from plantations. Local people have least incentive to prevent fires or keep grazing animals out of newly planted areas. In some countries, such as Philippines and Israel, government reforestation projects have been targets and anti-government forces, with divesting environmental impacts.

Brazil as to the highest rate of deforestation in the World.

The coastal forests of Ecuador, Ghanna, Madagascar have almost destroyed their forests. Haiti waste once 80 percent forested and today all the forests are destroyed and the land lies barren and eroded. About 2/3 of the forest in Central America has been destroyed, mostly in the past few decades as land is converted to pasture or cropland. Unfortunately the soil is poorly suited to grain of farming. A variety of reasons lead to this deforestation. In Seacoast Rica, in addition to cattle ranching, banana plantations consume large areas of forest. In Brazil, invasion of immigrants from South are responsible for forest losses. Gold mining in Ecuador, Coca production in Columbia and Bolivia and logging of hardwoods for export in Malaysia are some of the causes of forest destruction.

The other major source of forest destruction is invasion by land-hungry people from other areas. The loggers are interests only in creaming i.e., the most valuable wood such as teak, mahogany, San sood, or ebony. Although only one or two trees per hectare might be taken, widespread usually results. Because the canopy of topical forests is strongly linked by vines and interlocking branches, felling one tree can easily bring down a dozen others.

Reforestation

It means re-plantation of forest than is harvested. Many outstanding examples of successful reforestation are recorded. China for instance, cut down most of its forests one thousand years ago and suffered centuries of erosion and terrible floods as a consequence. Recently in China massive reforestation campaign has been started. Korea has also followed the policy of reforestation. After losing nearly all its trees during the Civil War, 30 years ago, Korea has now about 70 percent forested again. Japan has increased forests to approximately 68 percent in spite of being biggest importer of wood.

Satellite Fire Surveillance

Forest fire management is based on the concept that it is neither necessary not cost-effective to suppress fire that has occurred. Plenty of evidence suggest that most fires that are allowed to burn will be extinguished naturally before they have burned a few acres of trees. To assess occurrence of fire lookout towers have been constructed in remote forest regions. A constant watch was maintained during the summer fire seasons to detect fires as they occurred. Some of these towers are used, although fires are usually detected more effectively using modern technologies such as remote sensing equipment an Satellite Surveillance. Cost effective analysis are very important in suppressing forest fire. When a valuable timber resource is located in the area where fire is burning, it may be cost effective to suppress the fire.

Agroforestry

Conservation and public service organizations are encouraging agroforestry plans. These groups encourage people to plant fast-growing multipurpose trees such leucaena. Leucaen is a legume, so it fixes nitrogen and improves the soil. Millions of seedlings have been planted in hundreds of self-help projects in Asia. Africa and Latin America. Its nutritious leaves are good livestock fodder. It can grow up to 3 meters per year and quickly provide shades. They protect watersheds, create windbreaks and also provide useful good and forest products.

Selective cutting in place of clear- cutting

Clear cutting is a harvesting stem in which every tree is given area regardless of size. This method enables large machines to fell, trim and skid logs rapidly, but wastes many trees. This system increase soil erosion, and eliminates habitat for many forest species. Most timber and plywood in the United States and Canada currently is harvested by clear cutting. Size and shape of clear cuts vary depending on topography and management policies. In a forests managed for these trees, or forestires, blue berries, deer, or grouse, this is a good method. Other harvest practices substitute clear cutting with strip cutting. Strip cutting entails harvesting all the trees in a narrow corridor. The leas disruptive method is selective cutting. Under this method only a small percentage of the mature trees are taken in each 10 year or 20 year rotation.

Debe-for-Nature Swap

Developed countries of the world can save tropical forest from diminishing. This cab be done by financing nature. Finance is a great need for developing countries to protect and conserve forests. One promising approach to over power this problem is debt-for-nature swaps. Banks, government and money-leading institutions now hold nearly 1 million dollar loans to developing countries. This debe is not collected. Bank sell bonds at a steep discount nearly 10 percent on a dollar. Forest conservation organization buy debt obligation at a discount as mentioned above and then offer to cancel debt provided. Debtor country agree to protect, to restore an area of biological importance. The first such swap was made in 1987. Agreements have been reached with Madagascar and Zambia to swap debts for nature.

Forest Protection from Extinction

This has been explained earlier that tropical forests around the world are being cut at an alarming rate. Often hundreds of trees are smashed down to a single teak tree. This is seriously realized by tropical countries and they are initiating steps to protect them. People on the grass roots level have started to protect and restore forests. Indonesia has announced plans to preserve 100,000 sq. kilometres. Brazil and Zaire plan for forest protection in the world. School children plant trees in a community deforestation project in Haiti.

5

WATER RESOURCES

It is believed that life first orginated in water before it invaded land. Water is in faxt a pre-condition of life. Agriculture and growing urbanization with its modern life-style has been demanding greater share of water day by day. The primary source of water on earth is precipitation that comes in the form of rain and snowfall. The average annual rainfall in India is estimated at 117 cm. in a year. Water is available as surface water. The Indus, the Ganga and Brahmaputra carry 60 percent of total surface water.

As already described water is essential for all forms of human life. The surface of the earth measures, 50,000 million hectares in area, 20% of this area is land and the rest 80% is all covered by water. A man on a average consumes about 2 liters of water everyday. More than 70% of human's body weight is made up of water. Water as a natural resource is limited and estimated 1,011 mmckms of water on earth, out of which 486 mckms or 97% is contained in the oceans , out mckms is buried underground of the remaining small quantity of water 1.100th of 1% estimated at 33,400 m, provides the water supply vital for life on this earth.

In Delhi alone 800 million gallons is needed everyday but the treated water supply is 600 mgd, the short fall of 200 mgd is met by drawing groundwater. The increased demand for water can be met, if following water use methods are adopted because water is precious environmental resource. It is scarce, save it.

Following are the common methods of saving or conserving water :-

(i) Use a map and bucket of water for washing the floor instead of a running hose. One bucket of water contain nearly 18 litres will do the same work as a 10 minutes use of hose pipe. It contains 400 litres of water. We can save 382 litres of water.

(ii) Use a mug of water while brushing your teeth instead running tap.

(iii) Repairing immediately the water supply pipelines to prevent wastage and pollution of water.

(iv) Preventing pollution of water bodies. Once it is rained it takes years for replenishment.

(v) Registering all water-extraction points like tubewells and borewells.

(vi) Stopping use of treated water for gardening, washing vehicles.

(vii) Rainwater harvesting a techniques to recharge of groundwater by capturing and storing rainwater by constructing structures, such as dug wells, [ercolation pits, check dams. Rainwater is stored in groundwater reservoirs by adopting artificial recharge techniques to meet the household needs through storage in tanks, with the objectives: to reduce run-off, to avoid flooding of needs, to increase groundwater storage, and to supplement domestic water requirement during summer and long dry spells.

(viii) Adopt water harvesting techniques to recharge ground water levels.

(ix) Collect rainwater and use it for gardening and recharging ground water.

(x) Make sure that water does not overflow from your

overhead tank.

Multipurpose River Valley Projects

To achieve self-reliance and improver the standard of living of its people after independence, a number of river valley projects were developed. Many of them are multi-purpose river vary projects. These projects are meant to tackle various problems in an integrated manner. They control flood, check soil erosion, provide water for irrigation and drinking purposes, generate electricity for industries, villages and cities, provide inland navigation, and several other facilities like recreation, preservation of wildlife and development of fisheries. A brief review of the major multipurpose projects is given below:

(i) Rajashthan Canal Project : Rajasthan Canal Project is one of the important river valley project in India. It is one of the world's largest single irrigation project providing irrigation waters to nearly 19-49 lakh hectares in the districts of Bikaner, Ganganagar and Jaisalmer in Rajasthan. It entails construction of canal measuring 523.93 cubic metres capacity taking off the Harike barrage and the distribution system. Rajasthan Canal Project has been divided into two parts viz. (a) Rajasthan Canal Projecty Feeder and (b) Rajasthan Canal. Rajasthan canal is being fed by the supply of water by Ravi and Beas rivers. Later, the flow supplies will be supplemented by stored waters from Beas dam and Pong dams. The entire length of Canal and the feeder will be lined to save water losses through seepage. The intensity of irrigation will now be increased from the present 78 to 110%.

(ii) Farokka Barrage Project : The Farakka Barrage Project constructed over Bhagirathi Hoogly river system is designed for the preservation and maintenance of the

Kolkata Port. The river Bhagirathi, the feeder cabal and the Navigation locks at Farakka Barrage from part of the Haldia Allahabad Inland waterway. The main components of the project are (a) Navigation works such as locks, lock channels, shelter bosins, navigation lights and other infrastructure. (b) A feeder canal of 11333 cumec carrying capacity and 38.38 km. long lacking off from head regulator on the right side of the Farakka Barrage. (c) A 2,240 metre long barrage across the Ganga with rail-cum-road bridge, necessary river training works and a head regulator on the right side. The barrage is designed to pass a flood discharge of 76,455 cumec. (d) A 213 metre long barrage across the river Bhagirathi and Jangir. The barrage is designed to pass a discharge of 1,700 cumec.

(iii) Bhakra Nangal Project : Bhakra Dam, 225.55 m high straight gravity structure, is a majestic monument across river Sutlej. Its construction water taken up just after independence, for the uplift and welfare of the people of Northern Region and it was completed in the year 1963. The lake created by the dam is 168.35 sq. kilometers in area gross storage of 9868 million cu. metre the water stored at Bakhra have tremendous potential of generation hydro-electric power. Nangal Dam, 13 km. Downstream of Bhakra Dam, diverts the water of river Sutlej into Nangal Hydel Channel for power generation and irrigation purposes. Nangal Hydel Channel is a lined channel 64.5 km. long off-taking from Nangal Dam. Two power houses have been constructed on Nangal Hydel Channel; the first is located at Ganfuwal and second is situated at Kotla. The total installed capacity of Bhakra Power House and Ganguwal/Kotal Power Houses is 1354 kW.

(iv) Sardar Sarovar Project - Investigations reveal that

the potential of water resources and power development in the Narmada basin was initiated sometimes in 1947. The project was opened by the Prime Minister of India, with the availability of more detailed modernized counter sheets, raising the height of the dam for optimum utilization was considered and series of negotiation took place at the political level as it was an inter-state protect. The matter was subsequently given to the Narmada water Disputes tribunal as no agreement could be arrived at.

The government of Gujarat established Sardar Sarovar Narmada Nigam Limited as a wholly owned Government Company under the provisions of the Companies Ace, 1956. This is perhaps the first state level autonomous Corporation constituted for implementation of such a large multi-purpose project. The protect is an inter-state multi-purpose project of four states viz., Madhya Pradesh, Maharashtra, Gujarat and Rajasthan. Some of its main features are us under :-

(a) The River Bed Power House will be located underground on the right bank and will accommodate six revisable Francis Type turbine units, each of 200 MW capacity. The Canal Head Port House, also on the right bank, will accommodate five units of conventional Kalpan Type each of 50 MW capacity.

(b) The length of distribution system network will be about 75,000 kms. The supply of water would be volumetric by rotation with automatic remote control so as economies the use of costly and scare water.

(c) A 460 km. long main canal will be construct which will reach upto Gujarat-Rajasthan Border. It will have the capacity of 1133 cumes the largest in the world and it will taper down to about 71 cumes at Gujarat-Rajasthan Border.

(d) Creation of concrete gravity dam of 128 M. from the river-bed and reservoir, with live storage of 5800 million cubic mtrs. enveloping about 214km of the Narmada Valley and surface area of about 3700 hectares.

Benefits on Implementation of the Sardar Srovar Narmada Project

It is envisaged that following benefits would accrue on implementation of the Sardar Sarovar Projects :-

(i) Since it is proposed to compress the construction schedule of the project from 17/22 years to 10/12 years, on account of its early completion then a saving of about 3,500 crores on account of cost escalation would result.

(ii) On completion, annual additional agricultural production would be Rs. 900 crores. Power generation would be Rs. 400 crores and water supply would be Rs. 100 crores aggregation to about Rs. 1400 crores every year working out to about Rs. 4 crores a day.

(iii) It would also create employment potential of 7 lakh workers daily during construction period and 6 lakh workers daily during the post-construction stage.

(iv) Since new industrial corridor will develop around narmada main canal, Ahmedabad-Delhi meter gauge Railway line and Eastern highway would be the extension of the present rail-road corridor to New Delhi giving further boost to trade and commerce between Gujarat and Northern States.

(v) It will be a permanent solution for drinking water supply problem of water hungry regions of Saurashtra and Kutch.

(vi) Water supply requirements of several industries will also be met from the project giving all round boost of the industrial progress of Gujarat.

(vii) 4720 villages and 131 urban centres in Gujarat

State would get a permanent solution as regards the drinking water supply needs.

(viii) Power generation capacity of 1,450 MW would be added.

(ix) It would generate 5000 million units of electricity.

(x) In Gujarat State 17.92 lakh ha. of land spread over 12 districts, 62 talukas and 3393 villages, about 75% of which is drought prone, would come under irrigation every year.

Present Status of the Project :-

You will observe from the Directors Report that progress achieved in construction of the dam and appurtenant works, hydro - power, Narmada Main Canal and Branches as also distributary system have already been highlighted in detail would, therefore, like to dwell upon the novel and unconventional methods adopted by the Corporation for expediting the implementation of the project, the importance of which is as under :-

(i) The task of testing soil samples for main canal work has been entrusted to the private laboratories after short-listing.

(ii) Work of survey and investigation as also preparation of plans and estimates for the work of distribution system have been entrusted to private consultants.

(iii) To expedite the land acquisition for the canal works, the land acquisition organisation has been expanded.

It is learned that the approval of the Planning Commission for the Project was received in October, 1988. It is also informed that some inter-state key problems were resolved. The Government of Rajasthan has also indicated the level at which the Narmada main canal should deliver water on Gujarat-Rajasthan border. Moreover the issue regarding provision of irrigation bypass Tunnel and sharing cost of Rock-fill dykes and inter-linking channels is also resolve. The Government of India has given clearance to construction of approach road from Gora to Dam site which was very vital for jet speed concreting to Dam.

In order to expedite implementation of the Project to meet with the construction schedule which has been compressed to 17/22 years to 10/12 years, the Corporation has expanded the organizational base at all levels. For simultaneous completion of canal along with main canal levels. For simultaneous completion of canal along with main canal three posts of engineer-in-Chiefs and two post of Chief Engineers have been created. At present works of Rs. 1032 crores are on hand. The projects is completed.

Energy

Energy is a very essential need of a man. Any body can do an amount of work only when the energy possessed by the body is released. The sources of energy in rural system could be broadly divided into two ; traditional and modern sources. The traditional sources would be fuels wood, fuelstick, crop residue and cattle dung. These are partly commercial, conventional and renewable. Electricity, kerosene and other commercialized expect biogas, wind and ocean energy. The source of energy would depend upon the type of energy required for the purpose it has to serve.

Residential and Commercial Energy Demand

The amount of energy required for residential and commercial use varies greatly throughout the world. The economics of energy consumption describe that a country with high GNP product roughly use less per capita energy for residential and commercial needs, than a less developed country like Africa and India. 30 percent energy used in North America and 90 percent used in India is for residential and commercial purposes. Reason is very simple. North Americans use air conditioning, refrigeration, water heating and space heating. So 75 percent energy is used in these gadgets. In India, almost all of the energy used in home is for cooking due to scarcity and high cost of fuel. About half of the energy demand in Africa is for cooking because they use an open fire instead of stoves. Using more efficient stoves instead of open fires could reduce these energy requirements by 50 percent. Summarily North America, and India could reduce energy consumption in many ways.

Industrials Energy Demand. Considerable variation is found in the amount of energy used for industrial processes. Facts speak than 60 percent of all the commercial energy was used in the former Soviet Union. About 30 percent of energy is used by USA for industries.

In the former Soviet Union large deposits of coal, oil and natural gas reserves were available. On account of this reason steel was processed from ore because energy was readily available. Spain and Italy lack large deposits of fossil fuel necessary for the extraction of oil and natural gas. So both Italy and Spain produce steel from serape steel because it needs less energy. Producing 1 metric ton of steel from iron or in the former Soviet Union uses as much energy as producing a metric ton of steel from scrap in Spain or Italy.

6

ENERGY RESOURCES

Energy is an important input for development. No country can develop and progress in material pursuits without energy. There are broadly two sources of energy - Non conventional and conventional. Non-conventional energy sources are also called renewable sources of energy. Non- conventional or renewable sources are non-exhaustible. They are physical resources like coal, oil deposits, natural gas, minerals, soil, materials etc.

The other resources of energy are conventional or non- renewable. They are mostly biological in nature and include forestry, agriculture, animals. Solar energy, wind energy, water energy and geothemal energy belongs to this category. They are available in an exhaustible form in nature.

To keep pace with the increased demand of energy in the world both renewable and non-renewable resources are required to be harnessed fully besides conserving energy, because energy conserved is energy produced. For energy sources both non-renewable and renewable sources. Both Nonrenewable and Renewable sources of energy are discussed below turn by turn in an elaborate manner.

Wind Power or Energy

Moving air is called wind. This wind has energy. It is natural source of kinetic energy. It can pick up things and send them flying; it can lift heaps of sand and pile them into big sand dunes; it can hit the huge trees and send them crashing. This shows that the wind has a lot of energy to smash and destroy. But the man is intelligent enough to make this wind energy to do useful work for him. The energy possessed by the wind is due to its high speed or motion.

The possession of energy in wind is called kinetic energy. This kinetic energy in wind is utilized for doing work. In the earlier days, the energy of wind was utilized to sail boats for transport purposes and in wind-mills to pump out water from earth. A sail boat and a wind-mill is a machine which works with the energy of blowing air or wind.

There is abundant energy in wind which can be tapped. The World Meteorological Organization has estimated that 20 million MW of wind power could be commercially tapped worldwide. The Windmill is a proven technology. Windmills have been used for more than 1000 years, mostly to pump water and grind grain. Windmills were brought to Europe by merchants and returning crusaders, In the United States, the old farm windmill was used for decades to pump water and generate electricity. It is still used for electricity in remote areas for rural electrification.

Denmark and California leads the world in where it lacks other energy sources. The largest current market for windmills ins in the United States. Because of the immense amount of energy that could be produced by tapping the winds, private companies are building machines to tap the winds, then selling electricity to power companies.

Historically speaking windmills played a crucial role in setting of the American West. They had abundant water in underground aquifers but little water on the surface. The strong, steady winds that blew across the prairies, provides the energy to pump water that allowed agriculture to move west across the prairies. By the level of the nineteenth century, nearly every farm of Mississippi ever had at least one windmill. Even today some 150,000 windmills still spin productively in the Great plains. Manufacturing, installation and repair of windmills was a major industry in America at that time.

The General Electric Co. Estimated that wind could provide 14 percent of U.S. Electric demand by the year 2000. One thousand MW of installed wind power could meet the energy needs of 400,000 households at present use levels and save 3-7 millions barrels of oil per year. The General Electronic research unit have also discovered more than 556,400 sq./kg of land in he United States alone that are suitable for wind generators.

Coal

Coal is a major non-conventional source of energy used throughout the world. It is India's number one energy source. It is used for domestic cooking and industrial heating to the production of electricity, in thermal power plants. Coal is most plentiful fossil fuel. Coal is largely carbon mixed with incombustible mineral matter of earth's crust. When coal is burnt, it produces carbon dioxide and flash which is a particulate pollutant. It is a fossilized plant material preserved by burial in sediments and altered by geological forces. Coal is non-renewable source of energy because it takes very long time to form fossil fuel. Studies reveal that about 6000 billion tonnes of coal lies under the earth and by now around 200 billion tonnes had been used. Coal is among the most important traditional source of energy in India. 60,000 million tonnes of coal is minerable and sufficient to meet country's coal demand for another 130 years. Maximum coal deposits are found in Bihar and West Bengal in India.

Coal comes in various grades : anthracite, bituminous, subbituminous, and lignite in descending order of energy content and hardness. Anthracite, almost pure carbon, is found in very small quantities. An average bituminous coal might have an energy content of 30 MJ/kg while lignite may have an energy content of 16 MJ/kg. As with oil some coal is left in place when it is mined.

Though coal is mainly composed of carbon yet, it contains hydrogen, oxygen and varying amounts of nitrogen, sulphur and other elements. Coal was formed by the decomposition of the remains of vegetation growing in swamps. It is one of the most important source of energy and an essential input in steel and carbo-chemical industries.

Coal reserves are spread over almost 100 countries around the world. It contributes over 25% of world primary energy consumption. Coal is a competitive primary source for generation of electric power, World's 40% and India's 50% electric power generation is from coal.

India is the third largest coal producing country in the world. Per man productivity of coal is highest approximately to 2-45 tonnes. It is attractive investment regime having closely monitoring mine effluents with prescribing time limits for disposal of applications for mineral concessions.

The coal belts in India are located mainly in Orissa, Bihar, U.P., M.P., A.P., Maharashtra and Bengal. Coal as a source of energy is used for different purposes. Coal deposits are not evenly distributed throughout the world.

Notice that the countries with the largest land area have the most coal North America has one-fourth or all proven reserves. Antarctica is thought to have large coal deposits. But it is difficult, expensive and ecologically damaging to mine in Antarctica. Coal mining is dirty and dangerous business.

Oil

Oil is an equally important source of energy. Its use has been encouraged as being more convenient and often less expensive than coal. Its use became popular when automobile became quite common in 1920. This facilitated, convenient and less expensive use of oil compared to coal, stimulated exploration of oil. Today geologists use a series of test to locate underground formations that may contain oil. The moment, oil potential area is identified, a test well is drilled to determine if oil is actually present. Easy-to-reach oil fields have already been tapped. Drilling is being done in less accessible sites where smaller amount of oil is available.

This is a very expensive process of oil extraction. Moreover oil deposits in land have become more difficult to locate. Geologists are therefore occupied in search of location oil from the search of geologists. The cost of building an offshore drilling platform is very high and may cost trillions of rupees. To reduce the cost of drilling many wells may be sunk from a single offshore platform. Once the drilling platform is secured to the Ocean floor a number of wells are sunk to obtain oil and gas. This technique finally reduce the cost of constructing platform on the ocean floor. By the year 1993, estimated oil was extracted to the order of 1000 billion barrels. Of this, more than three quarters are in OPEC countries and half lie in Saudi Arabia, Iraq, Kuwait and Iran.

Oil exploration in India dates back to 1866 when oil exploration began Upper Assam and adjoining areas. Oil exploration made headway only after independence, more particularly over the past two decades. Only after independence, more particularly with a modest level of indigenous crude critical to India's economy. Starting with a modest level of indigenous crude production of only about 0.25 million tonnes in 1950-51, the production of domestic crude oil reached the peak level of 33 million tonnes in 1990-91. The production of crude oil during 1997-98 was approximately 33.86 million tonnes. The major contribution to the crude oil production has come only after the discovery of the offshore ail fields. Mumbai is the most prestigious among them.

The world's oil production during the past five years has a declining trend. There are two reasons for this. First is Iraqi invasion of Kuwait and secondly decline in Russian oil production. In Russia, the world's biggest oil producer, output fell form 12.4 million barrels daily in 1989 to 11.2 million barrels daily in 1992.

Facts About Oil : Indian Scenario

(i) Petroleum bearing state in India are Rajasthan, Krishna, Calvary, Godavery basin, Andaman, Bengal and Himalayan foothills.

(ii) ONGC, since 19567 is engaged in oil exploration, production and refining is done is 15 refineries with the capacity of 60 MMT per annum. Scottish exploration firm Cairn Energy has struck an estimated 20 billion tonnes of oil and gas reserves in the Barner district Rajasthan.

(iii) In India, total consumption of oil production is used in road transport sector amounting to 40% and the balance in power generation, domestic and miscellaneous purposes.

(iv) Petrol from plastic : Nagpur- based scientist Dr. Alka Umesh Zadgasankar claims it is possible to derive petrol from plastic waste. And with all the plastic waste choking our rivers and adorning our garbage heaps, a fuel from plastic is probably just what our environmentalists ordered.

(v) By 2005 Indian oil demand is expects to reach 2.8 million bad representing a growth of 6-6.5 percent annually from 2001, and about to reach 3.6 million bed by 2010. Crude oil imports are expected to reach 70 percent of domestic consumption by 2005. India's dependence on imported oil not only continues to increase but also makes the country one of the world's largest consumer of oil.

(vi) Oil reserves in the whole world are estimated to around 1,000 billion barrels. Keeping reserves and annual production, oil shall last for about 45-50 years. India shall have oil only for another 15-20 years because reserves are just limited to 785 million tonnes. India is planning to increase production of oil from 32.24 MT to 50 MT very shortly.

(vii) Hydrocarbon is crude oil and is converted form the remains of marine organisms due to heat, pressure and catalytic action. Conversion took millions of years.

(viii) Oil or petroleum is a complex mixture of hydrocarbons. It averages about two atoms of hydrogen to each carbon atom.

Energy from Biomass/ Bio-Energy and Biogas

Energy is being produced from biomass. Biomass resources used for producing energy include fuel wood, wood chips, bark, branches, leaves, starchy roots new plant growth, residues and wastes, herbage, algae, agricultural and forest residues like bagasse, corn cobs, bark, saw dust, wood shavings, roots and animal dropping etc. Sewage, garbage and night soil also provide biogas. Biodegradable organic effluents from industries like sugar, slaughter house, meat packing plants and distilleries are equally important sources of bio-energy. Biomass is renewable source of energy. Energy from this source is determined advantageous over nuclear and fossil fuels because of its renewability and easy accessibility. All renewable sources of energy account for about 18 percent of total world energy use, and biomass makes up three-quarters of that renewable energy supply.

Sufficient energy have been collected and stored in photo-synthetic organisms. They have been storing energy for the last 2 billion years. Studies reveal that plants capture nearly 0.1 percent of all solar energy that reaches the earth's surface. Some energy collected by the plants is spent in metabolic activities, the rest is stored in biomass. The metabolic activities are pumping water and oils, mechanical movement, maintenance of cells and tissues, and reproduction. It is difficult to measure the magnitude of energy, but experts estimate useful biomass production is of fifteen to twenty times, the amount of energy from all commercial sources of energy.

Biogas

Biogas is a sustainable source of renewable energy. Maintenance operation and construction of biogas units are simple. It provides multiple benefits to users. Biogas is a gaseous mixture of methane, carbondioxide and certain traces of nitrogen and hydrogen sulphide. It is composed of 60% methane 40% carbon dioxide. New Delhi has developed technology of briqetting saw dust into smokeless fuel- a good source of energy which yields biogas and pyrogas. Depending upon the content of carbon dioxide, biogas has calorific value of more than 5,000 kcal. The calorific value of biogas cab be further enhanced provided content of carbon dioxide is reduced and biogas is passed through a limestone solution.

Burning Biomass

Since time immemorial world fires have been the main source of heating and cooking. Till 1850 wood supplied 90 percent of the fuel to so called developed countries. But now wood provides less percent energy due to these nations because they have invented new sources of energy due to scientific and technological advancement. But the condition of underdeveloped countries is entirely different. In these countries, wood and other biomass fuels provide up to 95 percent of all energy used.

In northern industrialized countries of the world wood burning as a source of energy is very popular because oil, gas and coal prices are high. In these countries the supply of wood is also enough. It is believed that wood burning may limit further expansion due to hazardous amounts of carbon monoxide and hydrocarbons due to inefficient and incomplete burning of wood. In hill areas wood burning as source of energy is not recommended because it can present a major source of air quality degradation and health risk. Wood burning in these areas produce polycyclic aromatic compounds worrisome to human health because these compounds are carcinogenic. The Environmental protection Agency recommend use of wood stoves nationwide to overpower this problem. Clean-burning woodstoves are available but they are highly expensive.

Fuelwood crisis in underdeveloped countries in general and India in particular shall be faced because other sources of energy in these countries are not harnessed. Increasing wood use lead to diminishing fuel wood resources and has many adverse environmental impacts. As sallen wood becomes scarce, people lop off branches and even fell trees. These activities have following results:

(i) Air pollution resulting from burning biomass cause acid rain. The pH scale is used to measure acidity. A pH of 7 is neutral; when the pH is less than 7, the liquid is acidic; when pH is greater than 7, the liquid is alkaline. Lakes with pH less than 5.4 have no reproducing fish.

(ii) Contribute to decertification land climate change.

(iii) Cause soil erosion making deserts.

(iv) Depletion in ground water supplies.

(v) Destruction of forests and decrease in wildlife habitat.

The match between aggregate demand fuel wood and its supply would be highly disturbed in future. Projections reveal that by 2025, worldwide demand for fuelwood will be 4400 million cubic meters, while supplies shall not cross the present levels. Situation in African countries like Sudan and Rwada would be worst because their demand of fuelwood is already ten times more than supply of fuel wood.

Sources of Biogas Production

(i) Some forms of biomass energy are well suited to developing countries especially biogas digesters which are widespread in China and India.

(ii) Biogas is chiefly produced from : Cattle dung, human excreta, straw, plant, leaves, algae, bagasse, paddy husk, water weeds, distillery sludge, wastes from tannery, food industries and paper mills.

(iii) The anaerobic digestion process constitute

complex series of reactions catalysed by a mixed group of bacteria, 90% of the combustible energy of the degradable organic matter is converted into methane and carbon dioxide. The manorial value of organic waste is not reduced. The slurry from biogas plant is enriched manure having high degree contents of nitrogen, phosphorus and potassium. So biogas plants help in obtaining both fuel and manure from waste.

(iv) It is naturally occurring biotechnological process and begins when anaerobic decomposition of organic matter takes place.

CNG and LPG

Natural gas is mostly made up of methane. The other 5% is made up of various gases along with small amounts of water of water vapour. These other gases include butane, propane, ethane and other trace gases. Methane is a hydrocarbon, meaning its molecules are made up of hydrogen and carbon atoms. It is simple, one carbon, molecular structure makes possible its nearly complete combustion.

Like oil, this common fuel comes from underground. However, natural gas, as the name implies, is a gas much like air, rather than a liquid like petroleum. It has been found to be one of he most eco-friendly fuels, and its popularity is growing.

Because of its clean burning nature and the fact that it is not made from petroleum, as gasoline are, may automakers around the world are developing vehicles to run on natural gas. Cars, vans, buses and small trucks generally use natural gas that has been compressed and stored in high-pressure cylinders.

Several vehicles are available today that operate on CNG. Some run on natural gas only and others can run on natural gas or gasoline called bio-fuel vehicles.

Most people call liquefied petroleum gas (LPG) propane. That is because LPG is mostly made up of propane. Actually, LPG is made of a mixture of propane and other similar types of hydrocarbon gases. Different batches of LPG have slightly different amounts of the different kinds of hydrocarbon molecules. These hydrocarbons are gases at room temperature, but turn to liquid when they are compressed. LPG is stored in special tanks that keep it usually about 200 pounds per square inch. Most LPG produced in the U.S. comes from natural. Gas well head processing. That is because natural gas has LPG gases and water vapour in it, which have to be removed before the natural gas can be sent away in pipelines. Most of the LPG produced in caligonia comes from petroleum refining.

LPG is also used in many homes in the country, where there are no natural gas pipelines. These homes use LPG for heating, cooking, hot water and other energy needs.

LPG fuelled engines can pollute less than gasoline and diesel engines. LPG usually costs less than gasoline for the same amount of energy. In some countries LPG is used much more for vehicle fuel. In the Netherlands over 10% of the motor fuel used is LPG.

Ethanol and Methanol

Alcohol have been popular alternative fuels for many years. In fact, Henry Ford's first car was fuelled with alcohol. Both ethanol and methanol are now used as transportation fuels and will likely to play an increasingly important role in the future.

Methanol can be made from various biomass resources like wood, as well as from coal. However, today nearly all methanol is made from natural gas, because it is cheaper. Methanol is also very poisonous and very harmful if swallowed. Methanol must not be confused with ethanol. As with gasoline, it is also wise to avoid skin contact with methanol, as it can pass through the skin.

Sunshine or Solar Energy : A Bonanza of Renewable Energy

Solar energy is an important, clean cheap and abundantly available renewable energy. The sun reduates heat and light and provide energy to society. It is a source of enormous energy. Its approximate age 4 or 109 years and is expected to radial energy for another 4 × 10 years. Energy from the sun inn the shape of radiations is called solar energy. So you will be astonished to learn that with just 0.1% of the 75,000 trillion kWH of energy from sun reaching the earth is enough to fulfil the need of planet earth.

The solar cookers were put to good use in after the Gujarat earthquake. Credit goes to one of the professors of the Indian Institute of Management, Ahmedabad for this achievement.

The solar cooker is a lifetime asset. If handled with care it can last 15 years or more. There is no moving part in it to go wrong. The only maintenance required is an occasional coat of black paint on the inside of the cooker and on the outer surfaces of the cooking pots.

The solar cooker can be used both in the rural and urban sector. In the rural sector it saves village woman from having to trudge for miles in search of scraps of firewood and spending her lifetime in smoke filled kitchens to the great detriment to their health and also saves the messy cleaning of smoke blackened pots and pans. In the urban sector, it saves energy on kerosene and LPG and makes cooking easy. The cost of petro-based fuels is going up all the time by the gradual withdrawal of the subsidies. The price of a cylinder of LPG has already reached Rs 260 after the recent budged, and even this forbidding price has a subsidy element of 20%. This subsidy might also go away inn the coming months and years.

But the cost of solar cooker which is around Rs. 1500 is all the cost you have to bear and there is no recurring fuel cost. Even so the poorer sections may have to be helped with a subsidy or a facility of payment in installments.

Solar cookers are not being subsidized by the Indian government they were until 1994 though some states like Gujarat have continued the subsidy after 1994. But commercial fuels like kerosene and LPG, which are imported at great cost and add to the pollution, are subsidized. Even other renewable energy systems are subsidized, e.g. improved chulhas, biogas systems, PV systems, etc.

There is another class of cookers called concentrating cookers or direct-focussing cookers, which are concave mirrors which reflect solar energy to a focal point where a cooking pot can be placed. They are available inn several models. These are the SK 14 cookers and the Scheffler cookers. These cannot cook more than one item at a time but they can take a pressure cooker. They are large, unwidely and have to be fixed at a place and tare prone to scratches on the reflecting surface.

India may be justifiably proud that the world's largest solar steam cooking system is operation in the Brahmakumari's ashram at Mt. Abu in India. Here the solar energy is concentrated by a battery of concentrators to convert water into superheated steam. The system can cook for 10,000 people. This is truly a fascinating achievement of human endeavour.

Another type of solar cooker is also now being made in India, the Sunstove, designed by Richard Warehan U.S.A. It has been popularized in South Africa. It is an insulated box whose sloping sides themselves are reflecting - thus doing away with the outside reflector - and has a plexiglass window. It has two large sized pots. It is being make in Kolkata, with assistance form Rotary International. Disabled children of a Ramakrishna mission organization assemble the cookers. It has good prospects since it dispenses with glass components, is very light to carry and also cheap - costing around Rs. 1000.

Many individuals and organizations are working in several developing countries promoting cardboard panel cookers such as the Cooking which is simply a sheet of cardboard coated with aluminium foil cut and folded in a particular way so that it will reflect sun rays on a pot kept at its centre in a plastic bag which functions as a greenhouse. The are a good beginning because the are cheap and the people are too poor to afford anything better. But by their very nature they are not durable; do feel that one being made in India are required. The Cook-Kits also make interesting projects for boys and girls in schools. The Renewable Energy Policy, which is under the consideration of the government makes no mention of solar cooking or solar drying.

Solar Space Heating of Buildings

Architectural design of building occupies important position for providing solar space heating. The building in which heat is to be provided through solar energy should have large-facing windows. More radical possibilities include provision of an entire wall of double-glazed windows.

An alternative technology of solar space heating exits where heat is to be exchanged through water tank. It is necessarily an extension of the technology employed in solar water heating except that energy has to be recovered from the tank through a heat exchange surface. In this system, air is used as the heat transfer medium. There is no need of an additional heat exchanger as was done in the case with water. However, between the two i.e. water and air, water systems offer better heat-exchanger performance. Solar water system has to disadvantages such as leakage and doubt of water freezing. For inclement heating system using electricity or any other heating system is needed to back-up.

Solar Water Heating

This system of water heating is usually used in hospitals, hotels, guest-houses and hostels. A solar collector are of 3-4 m2 having insulated tank of 200-400 litre volume can provide 200-400 litres of hot water nearly at 600 C per average sunny day. The climate should be favourable. If more hot water is required than solar collector are with proportionate increase in the size of insulated tank. Following systems are to be designed for solar water heating unit.

(1) The collector pipes and tubes are connected to a hot water storage tank. The collector absorbs solar radiations and transfer the heat of suns to the circulation water. This is done either by gravity or by a pump. Storage tank is usually located above the top of the collection so that natural convection is obtained. Water circulates from the collected to the storage tank and no pump is required.

(2) Metal tubes are used for water to remove the heat usually provided with a glass cover. Layer of insulation

is applied beneath the plate.

(3) The collector is typical having a blackened metal plate absorber containing metal tubes or pipes.

(4) A flat plate collector and an insulated storage tank.

Solar Refrigeration

Solar refrigeration and air-conditioning are nearly same thing. Solar refrigerator is used for preservation of food, biological and medical materials. A number of experiments have been conducted by USA. France and Sri Lanka on Solar-operated coolers using absorption cooling cycles Solar refrigeration is usually used at household level for food refrigerators or small scale ice manufacturing units. Though machines and units are quite simple in design yet too complicated to operate. On account of this limitation its use has been dropped by many people.

Solar Air-Conditioning

Solar air-conditioning cover three different types of SAC as mentioned below:

(a) Solar regenerated Desiccant cooling system.

(b) Absorption refrigerator systems.

(c) Solar powered refrigeration systems of Rankine cycle systems.

Out of these, 2nd one i.e. Absorption Refrigerator system seems to have best prospects. In a desiccant cooling cycle, ambient air is adiabatically cooled, dehumidified, cooled evaporatively and sensibly as well, and then detected to the living area. Air is evaporatively cool and heated at the regenerative stage because it cools the supply air stream, heated again by solar collectors, and humidified. Dessicant cooling system is best suited for regions with about heating and cooling loads and high humidity.

Solar Heating and Cooling

Drying the body or warming in the sun during winters were perhaps two earliest uses of energy from the sun. But with the advancement of science and technology a number of varied uses of energy have been introduced by man. Now solar energy provide : Heat, Electricity, Mechanical energy, and Chemical fuels.

Solar Desalination

The idea of capturing sun energy or solar energy was developed and given practical shape in 1892 in Chile where solar desalination plant operated nearly 30 years. Solar energy is used for desalinating well water has been shown in many countries like USA. USSR, Greece in moderate sized pilot plants. Though the idea is old yet adapted to modern materials now in a modified manner. The process is given below :

(a() Take a transparent air tight cover of slopping sheets of glass.

(ii) Place sheets of glass into a shallow blackened basin or pool containing brine. Brine is saline water. Place it is the sun.

(iii) Solar radiation passes through the cover is absorbed and converted into heat.

(iv) This process cause water to evaporate from the brine.

(v) The vapours so produced get condensed to form purified water on the underside of cooler glass cover.

(vi) It flows down the sloping roof. Thereafter it is collected in the troughs and from there into a water storage tank.

(vii) The excess brine not evaporated in run to waste.

Per litre cost of this water is cheaper compared to distilled water obtained by other electrical energy-based processes.

Solar Drying

It is the most ancient and traditional method used for utilizing direct solar energy. Agricultural items such as copra, grain, hay, fruits and vegetables as well as animal products. Like spray of milk and fish are dried up all over the world where direct solar energy is available. The technique used involve spreading of the material to be dried up in thin layer on the ground in such a manner that it is exposed to sun and wind.

Recently some innovation have been developed particularly to fruit drying. To dry agriculture products in the solar energy, a box is arranged insulated at the base, painted black on the inner side and covered with the glass. Ventilation holes are provided both at the vase and top of the sides of the box to facilitate the flow of air over the drying material. The material which is to be dried is placed on well designed perforated trays or racks inside the cabinet carefully. The adoption of this technique has resulted product quality.

Studies reveal that a large portion of the world's supply of dry fruit, crops and vegetables continues to be prepared by sun drying. It is the cheapest and simplest method where we have abudant sunshine and post harvest is featured by low relative humidity and minor rainfall.

Solar Furnaces

It provides extremely high temperature form the sun under very clean conditions. The temperature generated and provide from sun is nearly up 350000 C. The heat so generated by solar furnaces is used for melting refractory materials. To obtain high temperature in a solar furnace following arrangements are designed :

(i) Turnable mirrors should be arranged on a slopping surface.

(ii) Concentrate the solar radiations on a number of turnable mirrors called the liostats.

Solar furnaces are used to study the properties of ceramics at extremely high temperature and the manufacture of fused aluminium crucibles. Solar furnace accomplish heating process without any contamination, and the temperature can be controlled by changing the position of the material in focus. It is a specialized device with least impact on the fuel market as a whole. Scientists and environmentalists opine that solar furnace shall be used in the production of nitric acid and fertilizers from the air.

Solar Greenhouse

The greenhouse is a solar collector which utilize solar energy for the growth of plants. It is closed structure covered with transparent materials made of glass or plastic. The short-wave incoming solar radiations can pass through the greenhouse glazing but the long-wave thermal radiations emitted by the objects cannot escape through the glazed surface. On account of this reason, the solar radiations get trapped within the greenhouse. This process bring increase in temperature. The air inside the greenhouse gets enriched with carbon dioxide. Due to transpiration, the moisture loss is reduced. All these happenings assist the plant growth to sustain during night and colder months of the year.

Geothermal Energy

Geo means earth and thermal means heat. Thus geothermal energy is the heat energy from hot rocks present inside the earth. This heat can be used as a source of energy to produce electricity. It is one of the few sources of energy that do not come directly or indirectly from sun's energy heat. It is heat of earth and may be defined as naturally occuring thermal energy fount within rocks formations and the fluids held within those formations. According to geologists the earth's core is a molten mass of material having vast amounts of energy. In the molten core of the earth, temperature are as high as 440000 C. The heat from the molten core of the earth offers an unending source of energy. Though the quantity of energy or heat stored in the earth is in abundance yet energy cab be exploited only in few particular areas called hot spots. Volcanic eruptions, gears and bubbling mud holes are symptoms of such hot spots. We have in India these spots at Vashisht bath and Manikaran. Both these spots are located at Manali and the other around 150 km away from Bhunter Tattapani near Shimla also provide hot spot.

Sources of geothermal energy

So far, following sources are known to us:

(i) Hot dry rocks found near areas of volcanic activity and energy is provided from water free, impermeable rock at high temperature and at practical drilling depth.

(ii) Geopressured brines hot water traced at pressures above the normal hydrostatic gradient. Geopressured brine contain dissolved methane.

(iii) Hydrothermal fluids naturally occuring hot sprints or vents.

(iv) Molten mass magma with temperature 12000C of material with very high temperature.

(v) Natural stream, hot waters or dry rocks.

WAVE ENERGY

Ocean wave energy has gained attention during recent years. Periodic to-and-fro, up and down motion of water particles in the form of progressive waves provide ocean wave energy. The period of ocean waves is of the order of a few seconds. It is an important renewable source of energy possessing both potential and kinetic energy. The ocean waves originate in different parts of the ocean surface due to the surface winds. The height of the waves depend upon the velocity of wind, depth of the ocean and contour of the shareline etc. Waves have typical ranges. Average wave height is worked out be 0.2 m to 4 m i.e. Wave period denoted by varies between 4s to occur. The may reach heights of 10 meters and may topple ships and damage the ocean energy plants.

Ocean wave energy can easily be seen and felt on any seashore. The energy that waves expect as millions of tons of water are picked up and hurled against the land, over and over, day after day, can far exceed the energy budget for both insulation and wind power in localized areas. Before providing details of wave energy, let us have a clear understanding of the term wave. Wave is long body of water curling into arched form and breaking on shore. Wind blowing over water generate waves. Ocean waves are derived from of solar energy. Wind act as an agent in transferring energy of the sun on the surface of sea. Ocean waves can travel vast oceanic distances with negligible loss of energy. Wave energy generated anywhere within an ocean basin finally reach to same island.

Environmental scientists estimated that only 0.01% of the solar energy flux is finally expressed as wave power. While sunlight falls on the earth at a global rate of 1016 to 1017 W, wave energy is dissipated against the world's coastlines at a global rate of 1012 to 1013

W. Coalines total 3,36,000 km in length. Andhra Pradesh is having longest coastline in India's.

Motion of waves have both vertical and horizontal movement of water. Particles of water undergo nearly in a circular motion individually. Individual particles move up as the crest approaches, become forward at the crest, down as it recedes and backward in the trough. Waves generate energy in their continual motion as detailed in the preceding line of the text. The energy so generated is harnessed either by a water operated or air operated turbine.

Air operated turbine is preferred to generate electricity. The torque of an incident wave causes relative motion between an absorbed and reaction point in all the wave energy conversion processes. The relative motion between Absorber and Reaction point drives working fluid through a generator prime mover. The characteristic which distinguish one process from another are (i) mode of energy absorption (ii) type of absorption and (iii) type of reaction point. There are 5 types of wave motions that could be used alone or in combination to provide energy.

TIDAL AND WAVE ENERGY

Extraction and transformation of force from the movement of water is called tidal power. The tidal movement of water represents a great deal of energy. This is called kinetic energy. All the flowing water carry this energy with them. The situation is like this that a flowing water encounters a turbine, a part of the momentum of the flowing water is transferred on to the turbine. This process make the turbine to rotate. The rotation of the turbine is used to generate electricity. This is how energy is produced from tides. It is an inexhaustible and renewable source of energy, available to society. Ocean tides contain enormous amounts of energy that is being harnessed to do useful work, since time immemorial. For years, engineers have suggested that moving water could produce electricity and the principle is the same as that employed in a hydroelectric plant. It makes no difference whether the water is in the open ocean, in a estuary or in a river. The estimated tidal power potential in India is about 15000 MW. The promising sites for tidal power, plants are located in Gujarat State and West Bengal State. Non-conventional and Renewable Energy Department is engaged in surveying other sites in Orissa, Tamil Nadu, Kerala, Karnataka, Maharashtra and Andaman Nicobar etc. These sites have good prospects but require extremely high investments compared to conventional power plants of same rating.

Based on experience, scientists state that the tides occurring in the oceans are one source of energy. Occurrence of tides in the ocean results due to centrifugal action and appearing or tending to fly-off gravitational forces. These actions and forces are caused by the rotation of earth as well as changes in the relative position of the earth, moon and sun. The process in called law of tides. As the size of moon size of the moon is relatively small compared to sun yet it is small in mid-ocean, but becomes amplified near coasts. The amplification is governed by the shape of coastline. The tidal range is of a bay or esturary leads to resonance effects.

The rise of ocean water to the attraction of moon is called high tide whereas the fall of ocean water is called low tide. The tidal waves in the oceans build and recede twice a day. The enormous movement of water between the high tides and low tides provides a very large source of energy in the coastal area of the world.

SOURCES OF GEOTHERMAL ENERGY

So far, following sources are known to us :

(i) Hot dry rocks.

(ii) Geopressured brines hot water traced at pressures above the normal hydrostatic gradient. Geopressured brines contain dissolved methane.

(iii) Hydrothermal fluids naturally occuring hot springs or vents.

Having Motion

It is vertical motion developed by Godforsaken. It has a buoy pulls in sea water to pump through the hose next time. The wave energy conversion process is depicted below :

(i) Surging - It represents horizontal motion in the direction of water travel. As waves get over a concrete barrier the fill a reservour. The water in the reservoir flows out through a turbine. One such device producing 350 kW was installed in Norway.

(ii) Pitching Motion - It was designed by Sir Christopher Cockerel. Pitch is an angular motion about an axis parallel to wave crest. It is a device known as Salter's duck which consists of a series of wing like objects connected together and appears to absorb most of the energy in waves. The system was determined too expensive.

Ocean Thermal Energy

The water at the surface of an ocean gets heated by the heat of the sun and attains a higher temperature than the colder water at deeper levels in the ocean. So, there is always a temperature differences between the water at the surface of ocean and a deeper levels. At many places in the ocean, this difference in temperatures of water is upto 200 C. The energy available due to the difference in the temperature of water at the surface of the ocean and at deeper levels is called ocean thermal energy. The ocean thermal energy can be converted into usable form of energy like electricity.

7

MINERAL RESOURCES

Resources obtained by mining are known as mineral resources. Mineralogy is science of minerals. Economic mineralogy is the study of minerals that are heavily used in manufacturing and are, therefore, an important part of domestic and international commerce. In nature, mineral are normally found in solid, liquid and gaseous states. India is rich in mineral resources. It has fairly abundant resources of iron ore and mica, and adequate supplies of magnesium ore, titanium, bauxite and coal. In all, there are over 3000 mines in India. About 8,00,000 people are employed in the mining sector. It accounts for about 11% of the country's industrial output. Public policy in the United States has encouraged mining on public lands as a way of boosting the economy and utilizing mineral resources. So is being done in India. World industry depends on about eighty minerals, some of which exist in plentiful supplies.

Three-fourth of the eighty minerals are abundant enough to meet all out anticipated needs or have readily available substitutes. At least eighteen minerals, including tin, platinum, gold, silver and lead are short in supply.

Of these 80 minerals, between one-half and one-third are considered strategic resources. Strategic minerals are those that a country uses but cannot produce itself. Government consider these minerals capable for military strength if unstable global economics or politics were cut off supplies. For this reason wealthy industrial nations stock-pile strategic minerals in times when prices are low and the supply in available.

Types of minerals : Geologists classify minerals according to their chemical composition and crystalline structure. Generally the are classified into tow categories. Metallic and Non- Metallic. Metallic minerals are further classified into two parts - ferrous and non-ferrous. Ferrous contain iron. There are iron ore, managenese ore, chromite, pyrite, tungsten, nickel and cobalt. Non-ferrous metals are gold, silver, copper, lead, bauxite, tin and magnesium. Not- metallic minerals are limestone, nitrate, potash, alolomite, mica, gypsum, coal and petroleum.

Rocks and their types

The crust of earth has many types of rocks. The material that forms the curst of the earth is called rock. Natural minerals are the ingredients of the said material. These minerals have natural occurrence with definite chemical composition. So rocks are the aggregate material of various types including harder materials like granite and soft materials like clay and sand. Rocks are divided into three major groups, explained below one by one.

Environmental Hazards from Mineral Extraction

Both mining and extraction of minerals have environmental effects. The main environmental hazards encountered are stated below :

(i) Water leaking into mine shafts also dissolves metals and other toxic material. When this water is pumped out or allowed to seep into groundwater aquifers, pollution occurs.

(ii) Deeper ore beds are extracted by strip or open pit mining where overlying material is removed by large earth-moving equipment. The resulting pits can be many kilometers across and hundreds of meter deep. Even deeper deposits are reached by underground tunneling, an extremely dangerous process for mine workers. In one of the coal mines in Chasnella 350 workers died a decade ago, when water entered while mining.

(iii) Mining destroys stream beds and fills the water with suspended solids the smoother aquatic life.

(iv) Uncontrollable fires, producing noxious smoke and gases, sometimes ignite coal-bearing scrap heaps stored inside or outside the mine. Surface deposits called tailings can cause acidic or otherwise toxic run-off when rainwater percolates through piles of stored material. Tailings from uranium mines give rise to wind scattering or radio-active waste.

(v) In coal mines natural gas poses dangers of explosion.

(vi) Old tunnels occasionally collapse or subside.

(vii) One of the most notorious effects of earthquakes is the tsunami. The seismic sea swells can move at 1000 km/hr, or faster away from the centre of an earthquake.

(viii) Earthquakes and volcanic events are natural geological hazards that are result of movements of the earth, restless core and mantle.

(ix) Smelting and chemical extraction processes also create pollution problems.

(x) Surface mining results in removal of natural eco-systems, soil disruption, creation of trenches and open pits.

(xi) Fumes form smelters kill forest and spread pollution over large areas.

(xii) Mine drainage pollute thousands of kilometers

of streams and rivers.

Sedimentary Rocks

Sedimentary rocks are also called stratified rocks. These rocks are formed by the sediments or deposits laid down on the floor of the river beds, lakes or seas. Sedimentary rocks contain different types of dediments like sand, gravel, limestone, slate and chalk. History of geography reveals that chalk has interesting origin. The chalk is made up of skeletons of the tiny creatures that lived in ancient seas. When the creatures died, their skeleton fell to the bottom and over million of years layer of skeletons piled up which ultimately got squeezed into hard mass. So sedimentary rocks have enough contents of chalks. A distinguishing feature of sedimentary rocks is that they contain fossils. The fossils are greatly helpful inn estimating the age of the sedimentary rocks. Coal and limestone are the best examples of sedimentary rocks.

Igneous Rocks

These rocks are called the primary rocks because these were found first on the surface of the earth. These rocks are formed by cooling and solidification of the molten magma. Magma gets erupted by volcanoes. Basalt and granite are examples of igneous rocks.

Limestone

1. Country's production comes from Madhya Pradesh, Chhattisgarh, Andhra Pradesh, Rajasthan, Gujarat, Karnataka and Himachal.

2. Found in geological formations because of having sedimentary origin.

3. 75% used in cement industry and rest for smelting of iron.

4. Found associated with rocks composed oı calcium

carbonates.

Mica

1. Exported to U.K., U.S.A., Russia, and exported through Kolkata and Vishakhapatnam.

2. India produces 60% of the world's production of mica.

3. Occurs in Jharkhand, Bihar, Andhra Pradesh, and Rajasthan.

4. Used in electrical and electronic industries having quality to withstand high voltage.

5. It has insulating properties.

Bauxite

1. Occurs mainly in Jharkhand, Orissa, Gujarat, Maharashtra, Chhattisgarh, Madhya Pradesh and Tamil Nadu.

2. Widely distributed in India with vast reserves.

3. Used for manufacturing of aeroplanes and utensils.

4. Aluminium is obtained form bauxite.

Lead

1. Available at Andhra Pradesh, Rajasthan and Tamil Nadu.

2. Produced 25% of the requirement.

3. Imported from Australia, Canada and Myammar.

4. Used as constituted in alloy. Lead oxide is used in cable covers paints, glass, and rubber.

5. The ore of lead is called galena. It is soft and heavy metal./

Manganese Ore

1. Available in Karnataka, Orissa, Madhya Pradesh, Maharashtra where we have 97% mines of manganese.

2. Stands 2nd in the reserve of manganese ore in the world after Zimbabwe.

3. 1/5th exported. Only low grade manganese is exported to Japan. Which buys 2/3 of India's export.

4. 4/5 the of the production is consumed in the

country/

5. Also used to manufacture bleaching powder, insecticides, paints and batteries.

6. Used for making iron and steel, alloys etc.

Iron Ore

1. Available in Jharkhand, Chhattisgarh, Andhra Pradesh, Goa, Orissa, Karnataka, Tamil Nadu, Maharashtra and Rajasthan.

2. Iron ore is available inn 4 varieties : magnetite, Limonite, hematite and siderite.

3. Exported mainly to Japan through ports of Vishakhapatnam, Paradip, Marmagoa and Mangalore.

4. India's share is 20% of the world.

5. Used for manufacturing of machines and implements.

6. Backbone of modern civilization a metal of universal use.

8

SOIL RESOURCES

Soil is a thin covering over the land. It is marvelous substance a living resources of astonishing beauty. Soil is an organised and complex mixture of weathered mineral materials from rocks, living organisms, air, water and partially decomposed organic molecules. Soil is an essential component of the biosphere and must be natured and cultivated to bring to its highest potential. There are a least 20,000 different types of soils in the United States alone and many be hundreds of thousands throughout the world. Survival and growth of rural and urban population is governed by soil and its nature. In rural areas farmers are particularly concerned with soil because the nature of soil determines what kind of crops can be grown and what cultivation technique should be employed.

Urban dwellers are also concerned with the nature of soil because the health of soil determines quantity and quality of food they will eat. To understand the potential for feeding the world on a sustainable basis, we need to know the formation, composition as well as how soil is being lost and what can be done to protect and rebuild good agriculture soil.

Soil Profiles

A cross-sectional view of the horizons in a soil is called soil profiles. Classification of soil as per its thickness, colour, texture and composition determines soil profiles. The soil profile is a series of horizontal layers of different chemical composition, particle size, and amount of organic matter. Each recognizable layer is known as horizon. The different layers of soil differ physically, chemically and biologically. These layers are classified into different horizons. These horizons are three in order as per their profiles.

A - horizon : This is the top layer of soil. It contains most of the organic matter. The thickness of this layer of A-horizon may vary from less than a centimeter on steep mountain, slopes to over a meter in the rich grasslands. It is dark-coloured horizon with much organic matter mixed with mineral matter. The lower portion of the A-horizon often contain, few nutrients because water flowing through the soil from the top to deeper layers dissolves and transports nutrients to the B-horizon. This process is known as leaching.

B-horizon : This horizon is called sub-soil. It contains less organic material and few organisms. It has accumulated nutrients that were leached from higher levels i.e. A horizon. B- horizon is very valuable source of nutrients for plants. It also support root system of plants.

C- horizon : The area below sub-soil or B- horizon is known as the C-horizon. It contains weathered parent material. This parent material contains no organic materials but it does contribute to some of the soil's properties. The chemical composition of this soil helps to determine the pH of the soil. C- horizon soil also influence the soil's rate of water absorption and retention.

Factors contributing to soil development are extremely varied. Nearly fifteen thousand separate soil types have been classified in North America. In other countries of the world, there could be more types of soil. However many of the soils can be classifies into two broad categories isle. grassland soil and forest soil.

Grassland Soils : Soils formed in grasslands are known as chernozem soils and have deep A-horizon. The low amount of rainfall in grassland areas limits the amount of leaching from the A- horizon or topsoil. This lack of leaching also results in a thin layer of sub-soil.

Forest Soils : These soils develop in areas of more abudant rainfall. The rainfall results in a layer of topsoil that is usually not as thick as grassland soil. But the material leached from the topsoil forms a sub-soil which support substantial root growth.

Soil Properties

The properties of soil are explained below :

(i) Soil structure : Soil structure refers to the way various soil particles clump together. Sandy soils lack structure, whereas clay soil tend to stick together in large culpam. A good soil forms small clumps. Due to small clumps this soil crumble easily. The ability to crumble is called soil's friability. This characteristic of soil is determined by the soil structure and its moisture contents. Sandy soils are very friable. Clay soils do not have friability. A good soil will crumble because of has spaces to allow air and water to mix with the soil. The contents of air and water depend upon these spaces. In good soil, about 2/3 of the spaces contain air after the excess water has drained. The air in these spaces provides a source of oxygen necessary for plant cells. Water for the roots occupies the remaining soil space. The relationship between the amount of air and water is not fixed.

There are some typical inhabitants of soil stated below:

1. Bacteria : Bacteria and fungi are important in the

decay and recycling of materials. Their chemical activities change complex organic materials into nutrients that can be used by plants. These micro organism can convert the nitrogen content in the protein contained component of organic matter into ammonia or nitrate, nitrogen compound that can be utilized by plants.

Soil organisms usually stay close to the surface, the thin living layer can contain thousands of specials and billions of individual organisms per hectare. Algae live on the surface while bacteria and fungi flourish at the top, few centimeters of soil. A single gram of soil can contain hundreds of millions of these microscopic cells. Round worms, segmented worms, mites and tiny insects appear in thousands in that same gram soil.

2. Soil Texture : Soil contain rock particles. The size of rock particles within the soil determine soil texture. The largest size of rock particle in the soil is called gravel. Gravel consists of fragments larger that 2-0 millimeters in diameter. Particles between the size of 0-05 and 2-0 millimeters are classified and sand. 0.002 to 0.05 millimeters in diameter refer to silt particles. Smallest particles within soil are clay particles which are less than 0/002 millimeters in diameter.

There are many tiny spaces between large particles such as gravel and sand. The spaces between the two that is sand and gravel allow both air and water to flow through the soil. Water drains from this king of soil vary swiftly and often carry valuable nutrients to lower soil layers, where they are beyond the reach of plant roots. Clay particles being flat are conveniently packed together and form waterproof layers. Since water does not flow through clay very well, clay soils tend to moist and do not easily low minerals to flowing water. Soil consist of many sizes of particles. But an ideal soil is a loam.

Soil Composition and Formation

We know soil nutrients are depleted due to many farming techniques and exposure of soil to the erosive forces of wind and moving water. The lost nutrients of soil can be reestablished and renewed with careful husbandry. This soil becomes young by proper care and provide greater and greater productivity.

Building good soil is slow process. Good topsoil accumulates 10 tons per hectare per year under good circumstances and favourable environmental conditions. Under poor conditions, it takes thousands of years to build good soil. Topsoil is loosing its fertility much more faster than it is being replaced. Soil experts state that one third to one-half of the world's croplands are loosing topsoil faster than it is being renewed. In some parts of the world, erosion carries away about 2.5 cm of topsoil per year.

The formation of soil is a combination of physical and biological events. Soil building begins with the physical and biological events. Soil building begins with the physical fragmentation of the parent material. Parent material consists of ancient layers of rock as well as recent deposits from lava flow. It is the parent material and the climate that determine the kind of soil that shall be formed. Weathering bring about fragmentation of rocks. Large rocks get fractured the moment they are heated because rock does not expand evenly. Pieces of the rock flake off and are often reduced into tiny sizes by other processes. These processes are repeated freezing and thawing of water. Water that has seeped into cracks of rocks expands as it freezes. This process cause cracks in the rock to widen the crack in the rock fills with water.

As the water freezes and becomes ice, the pressure of the ice enlarges the cracks. The ice melts, and water again fills the rock. The water freezes again and widens the crack. Alternating freezing and thawing splits the rocks into smaller fragments.

Physical breakdown of rock also appear due to forces that cause rock particles to move and rub against each other. In many parts of the world, the parent material from which soil is formed consists of glacial deposits. Wind and moving water also cause small particles of fragmented rocks to collide. This process also assists in soil formation. The smoothness of rocks as well as pebbles seen on the shore line and river beds is evidence of this rubbing process.

Soil building process is also done through chemical weathering Small rock fragments when exposed to the atmosphere may be oxidized; that is, they combine with oxygen from the air and chemically change to different compounds. Other kind of rocks may combine with water molecules in a process known as hydrolysis. Hydrolyzed molecules are more readily soluble in water and can be easily removed by rain or moving water. Rain has slightly acid content which assist in dissolving rocks.

There are certain organisms which trap the modified parent material that contribute to the process of fragmentation. Lichens form an important community that trap small particles of fragmented rocks. These organisms chemically alters the underlying fragmented rocks. Other kinds of organisms such as plants and tiny animals become established and contribute increasing amounts of organic matter which are included with the fragments of rocks to form soil. Decaying organic matter called humus becomes mixed up with top layers of rock particles some nutrients needed by plants and also increase the acidity of Soil. Wheat and corn. Soils with the best texture for most crops are looms. If soil were 40% sand, 40% clay, it would be loam.

Burrowing animals like earthworm as well as roots of plants are equally important organisms that assist in the process of soil formation. One hectare of soil may support a population of five hundred thousand earthworms that can process as much as 9 metric tons of soil a year. Burrowing animals bring nutrients from the deeper layer of the soil up into the area where plant roots are more concentrated thereby improving the soil's fertility. Fungi and bacteria are important decomposers in improving the quality of soil. In certain countries of the world geologic forces like glacier, wind and water contribute enough to the parent material upon which soil composition depend. About 97% of all the parent material in the United States was transported to its present site by geologic forces.

Reasons and Mechanisms of Soil Erosion

A few mechanisms are involved in the process of soil erosion. These processes of soil erosion are recognized and called as sheet erosion, rill erosion, gully erosion and stream bank erosion. There could be other types of erosion too, but fact remains that wind and water are the main agents that move soil around. However, the type of erosion as mentioned in the preceding lines are explained below in a nutshell manner.

(i) Streambank Erosion : This erosion is caused by washing away the soil from the banks of established streams or rivers. Soil is washed as a result of removing trees and brush along stream-banks. Most soil erosion on agricultural land is sheet and rill erosion. Large amounts of soil is being transported by this mechanism without being noticed. A farm field can lose 20 metric tons of soil per hectare.

(ii) Gully Erosion : When rills enlarge to form bigger channels or ravines difficult to be removed by normal tillage operations, then this process is called gully erosion. Gully erosion is an advanced stage of rill erosion. Any concentration of surface runoff is a potential source of fulling. Unattended rills get deepened and widened every year and begin to attain the form of gullies. During every rain, the rain water rushes down these gullies, increasing their depth, width and length. Within a few year's the entire landscape may be filled with a network of gullies. Large areas of fertile lands thus get swallowed by the ever-increasing gullies and are rendered uncultivable.

(iii) Rill Erosion : The process is called rill erosion when little rivulets of running water gather and cut small channels running down in the soil. Rills develop when there is a concentration of fun off water which, if neglected, grow into large gullies.

(iv) Sheet Erosion : This erosion occurs when thin uniform layers are peeled of the land surface by water and wind. It is caused by rainfall and water run off. This type of erosion is extremely harmful to the land and usually show that the farmer is not conscious of its existence. It is common on land having gentle slope.

Causes and Reasons

Water and wind are important mover of soil. How water causes soil erosion has already been explained. In this section it is proposed to discuss how dust and wind, particularly dust bowls is responsible for replenishing soil. Wind erosion may not be as evident as water erosion, since it does not have gullies on the surface of the soil but erosion caused by water is quite evident. So the main causes of soil erosion as experienced from time to time by environmentalists and soil scientists in different countries of the world are discussed below :

(i) Agricultural Machinery : Farmland is being manages because agricultural machinery is becoming heavier, more powerful and more widely used, says a survey. The soil is the loser. It is increasingly compressed by tractors, combing harvesters and other machinery. The quality of farmland can decline markedly as a result. Inadequate nutrient supply used to be mainly to blame for low yields ; it is now often joined by damage caused by the soil being compacted. The survey, carried out for the Bavarian Environment Ministry by Rainer Horn, who is a Bayreuth University soil physicist, describes what happens when the soil is churned up by tractor types. Soil, he says, has a natural support, a structure consisting of expanded clay aggregate, clay, sand and humus. This substratum expands and contracts in the soil, forming a grid that supports the topsoil and ensuring sound soil structure. Healthy soil can withstand the pressure of tractor tyres and has sufficient bent to spring back into shape. Any compacting that occurs can at least be offset in part.

These stabilizing aggregates cannot withstand unlimited pressure. There is a limit to the stress a given variety of soil can withstand. If it is exceeded the aggregate is compacted, water is pressed out of the topsoil and soil is rutted for good. Deep ruts and churned-up soil are unmistakable danger sings. When this point is reached will depend mainly on the variety of soil and on its humidity. Heavy, clay soils are much more sensitive than sandy soil. The wetter soil is, the more readily it can be compacted. The reason for this could hardly be simpler. Soft and soggy aggregate is less firm than the hard, dry variety. The consequences for crop plants are serous. Soil ventilation deteriorates, leaving roots starved of oxygen and water less easily accessible. Cohesive forces in small-pore soil than they would in more loosely packed earth, which can lead to lower yields.

The growing threat of erosion is a further problem. Once the system of cracks and fissures in the soil has been destroyed, rainwater can no longer penetrate fast enough. It is drained off superficially and carries off topsoil with it. Compacting of soil is very difficult to reverse. Attempts to plough deeper and vermiculate it better have often failed, merely transferring the compaction to a lower stratum. Horn feels nature's self-recuperative capacity is the sole hope of recovery, nature in this case consist of a combination of the earthworm and ground frost.

Ice the freezes in the soil expands and can thus open up comacted earth. But even in fallow fields it can take years for the soil to be regenerative. So the aim must be, as far as possible, to prevent overexposure to stain in the first place. Horn has experimented with vehicles, drinking them over ground in which sensors that respond the pressure have been driven. He has found that soil is damaged less low-weight tractors crossing frequently than by juggernauts churning up the soil less often. A heavier vehicle weight, it was long felt, could be offset by a larger type surface area. But the Bayreuth readings disprove this assumption.

Type pressure spreads through the soil in what Horn describes as onion fashion. The narrow tyres of a light-weight tractor produce a correspondingly small onion. The jumbo tyres of a larger agricultural machine may generate the same pressure per square centimetre but the larger surface area makes the pressure onion expand. As a result the stain is sent deeper into the ground and the soil is impacted even further. So the only way to reduce the pressure is to fit larger tyres to smaller items of machinery.

(iii) Modern Agriculture : Organic decline in soil cause soil erosion due to the use of modern agriculture techniques. Two recent changes in farming techniques have helped to reduce the amount of organic matter in soil. First, farmers have ploughed up many old permanent pastures to plant crops. In other words, pastures growth is abandoned and in its place plant crop inn encouraged. Pasture crops maintain or even raise the amount of organic matter in soil, whereas arable cropping tends to reduce those levels. Secondly, inorganic fertilizers have largely replaced organic manure. Large applications of inorganic fertilizers can contribute organic matter to the soil by increasing the yields of crops.

However, where the land is continuously under arable crops, or where the residues of crops are removed or burnt rather than being ploughed back into the soil, the organic matter is lost more rapidly than it is replaced. Grass crops not only increase the amount of organic matter, but they also permanently cover the ground, affording greater protection to erosion by rain. Modern farming also famous large fields because they are quite suitable for modern machinery. But then large fields increase wind erosion, because hedges and shelterbelts slow the wind at ground level and decrease the distance of uninterrupted blow.

(iii) Dust Bowls : They are also called dust storms. These are worst human- caused environmental disaster ever experienced in United States, 10 billion tons topsoil was moved by these dust bowls in 1935. Thousands of people died of dust pneumonia while millions join the mass migration. A prolonged drought beginning in 1931 was the immediate cause of the dust storms had reached $ 25 per day. It seems incredible that something as basic as the very soil inn which we stand should be disappearing at such a rate that by the end of the century there will be 32 percent less person than at resent. And yet there it is. A fact that is a obvious as any to those who have seen the dust bowls of the American prairies. The spectre a now dust bowl and worldwide food shortages was evoked by US ecologists as dust clouds hundreds of miles wide swept by 60 mph winds from the drought-parched soil.

The demands made upon the Earth's agricultural are becoming greater, but the answer does not lie in the use of more and more chemical fertilizers. Neither does it lie in ploughing up more and more marginal land, or shortening cultivation cycles. And yet another stark fact on our list is that we have an increasing number of people to feed, and, just as important, a considerable and wealthy minority of them want to eat the type of foods that wasteful in the amount of energy needed to grow them. How do we reconcile this with the problem of erosion apparently caused by the practices that aim to alleviate hunger, and the rich world's greed?

Calculations suggest that by the turn of the century there will be third less topsoil throughout the world per person than at present. Almost half of American's farmland continues to lose soil faster than it is replaced : in the early 1980s, 1700 million tonnes of soil were lost each year. Where does all this eroded soil end up is note worthy point. Some of it is blown across land and out to sea. It shows up on satellite pictures or it is monitored by stations on islands that take air samples. Other quantities of eroded soil end up being washed out to sea or silting up dams and reservoirs considerable reducing their useful lives.

Soil erosion history

There is close relation between history of soil erosion and the history of agriculture. Expletive agriculture accompanied by the cutting of forests on the attachment of rivers began at the dawn of history. Soil erosion has been responsible for the fall of various ancient civilizations Mesopotamia, and land well developed in the time of the Babylonians, Assyrians and Persians in now virtually desert. In India, the Mohenjodaro and Harrappa civilization became extinct because of neglect of their lands. Rivers like the Saraswati have disappeared as their watersheds and river basins were exploited and mis-managed. A spectacular care of erosion in that of Punjab Shivaliks, a range of hills fringing the Himalayas. Along most rivers in India, soil erosion has led to the formation of a vest network of fissures and gullies interfering with the use of agriculture land. There is ravine land adjoining most of the Indian rivers, but Chambal is the worst offender in the respect, where several thousand hectares of land have been destroyed. Even in the great alluvial plains of India where the land is comparatively flat, the damage by gullying is considerable.

Soil Conservation Practices

Sustainable agricultural of regenerative farming depends upon careful husbandry of soil. The most productive layer of the soil is topsoil which is usually removed and lost by water or wind erosion. This has adverse effect upon agriculture output because agriculture is the area in which topsoil is most essential.

When the topsoil is lost, the soil's fertility decreases. To re-establish, this lost fertility fertilizers are used. This has the effect of raising the cost of the food we buy. Besides this, excessive movement of soil from farmland to streams have many undesirable effects as discussed below one by one.

(i) The soil carried by a river is eventually deposited somewhere and in many cases it should be removed to clear shipping channels.

(ii) A stream laden with sediment affects the fish populations. Fishing may be poor because of unwise farming practices hundreds of kilometres upstream.

(iii) Eroded and moved soil make streams dirty and muddy. So a dirty stream is less aesthetically pleasing than a clear stream.

In view of these reasons, proper soil conservation measures are required to be adopted to save the topsoil being eroded. These measures can be as detailed below:

(1) Windbreaks - In a relatively dry areas of the world soil erosion caused by wind is a serious problem. Wind erosion can be checked if the soil is protected from the wind. This can be done by planting generation over the surface of the soil. Soil is yet exposed to wind in spite of this practice. So the better way is to plant trees that protect bare soil from the full force of the wind. Windbreaks reduce the velocity of the wind, thereby decreasing the amount of soil that is carried away. In some cases, rows of trees are planted at right angles to the prevailing winds to reduce the force of the wind, while in other cases, kind of strip farming is practiced.

(2) Waterways - Though contour farming, strip farming and terracing are quite effective approaches directed towards soil conservation yet cannot provide wonderful results without the practice of waterways.

Waterways are depressions on sloping land where water collects and flows off the land. These areas are exposed to erosion when not properly maintained. If a waterway is maintained with permanent sod covering, the speed of the water is reduced, and soil erosion is decreased.

(3) Terracing - The terraces are very important for people who live in countries that have little flat land available. Terracing involves shaping the land to create level shelves of earth to hold water and soil. The edges of terraces are planted with soil-anchoring plant species. This agriculture practice though useful in controlling soil erosion yet quite expensive needing much energy and hand labour. Terrace farming use of land without serious erosion.

(4) Strip Farming - Strip farming consists of alternating strips of closely sown crops like hay, wheat or other small grains with strips of row crops like corn, soya beans and cotton. This method of preventing soil erosion is very effective where a slope is too steep or too long. Contour farming and strip farming together is used for raising crops when slope is too steep or too long.

(5) Contour Farming - This is the simplest method of preventing soil erosion. You know water runs downhill. The faster is runs, the more soil it carries off the fields. This is called run-off. Water run-off can be reduced by contour ploughing. This practice is useful on gentle slopes and produces a series of small ridges at right angles to the slope. This practice allow more water to be soaked into the soil. Contour farming reduces soil erosion by as much as 50 percent. Contour farming increases crop yields by conserving water. The soil conservation practice is useful on gentle slopes.

Soil Conservation Methods

Soil conservation problems vary with geographical

locations, soil types, land use watershed characteristics and climate. Major soil erosion problems and how these special problems can be solved with the application of remedial measures are discussed below.

Modern Farming

Modern farming favour soil conservation by checking soil erosion. This is a known fact that large ficlds enhance wind erosion, because hedges and shelterbelts slow the wind at ground level and decrease the distance of uninterrupted blow Hedges and ditches also play a similar role on sloping land that is prone to erosion by water. They break up the length of the slope, which slows the water running off and thus reduces its capacity to erode the soil.

Some sample measures can limit erosion without drastically altering current farming practices, farmers can grow grass of steep slopes, avoid cultivating up and down shallower slopes, and replace hedges and ditches on slopping grounds. These changes, however, will no stop the loss of organic matter in many soil is nor prevent damage from heavy machinery when the ground is wet.

Sea Erosion

Sea erosion is caused during rainy conservation by checking soil erosion. This is a known fact that large fields enhance wind erosion, because hedges and shelterbelts slow the wind at ground level and decrease the distance of uninterrupted blow Hedges and ditches also play a similar role on slopping land that is prone to erosion by water. They break up the length of the slope, which slows the water running off and thus reduces its capacity to erode the soil.

Some sample measures can limit erosion without drastically altering current farming practices, farmers can grow grass or steep slopes, avoid cultivating up and down shallower slopes, and replace hedges and ditches on slopping grounds. These changes, however, will not stop the loss of organic matter in many soils nor prevent damage from heavy machinery when the ground is wet.

Land slips and slides

This type erosion is usually seen in the hills of Himalayan region. Land slides caused to erosion block high ways and disturb the communication system. It destroys buildings constructed below as well as silt reservoirs, Land slips and slides are caused by the hydraulic pressure developed by the moisture penetrated in the soil to a saturation point during heavy rains.

The main causes of land-slides are under cutting of the base of the exiting slope, disintegration of soil through development of cracks and insufficient drainage in the sliding mass.

Remedial measure used are given below :

(i) Stabilization of small and big streams in the watershed in done by constructing a series of check dams.

(ii) Slipes are broken by contour planting to care the growth of vegetation.

(iii) Use logwood crib chen dams to cover the rills on landslide face.

Road side erosion

Some king of erosion take place along the roadsides and rail lines. This type of erosion is mostly found in the hills where the roads have been constructed by the cut and fill techniques and where the batters protected. On account of this reason, the situation gets aggravated when drainage in the area gets adversely affected.

The methods of checking roadside erosion are given below :

(i) In order to over power the adverse effects of meandering rivers on the railway lines guide bunds with pitched slops, are made.

(ii) Provide roadside drain for carrying the run off from the fill slopes.

(iii) To control the run-off from the cut-face catch water drains or gutters proper outlets should be constructed.

(iv) Turning, pitching and construction of retaining walls.

(v) Constructing interceptor drains about 3 to 5 meters away on the upstream side form the top of the cut face.

Stream bank erosion

Stream bank erosion takes place when the runoff flows over the side of the banks are devoid of vegetation due to overglazing. Velocity of the flow in the stream is considerable influences by the degree of erosion. Any obstruction to the stream flow forces the change in the stream course which cause stream bank erosion. When the stream or river gets loaded with sediments carried by the runoff the course of the stream gets shifted and in the process the stream banks get eroded. This is predominant in case of rivers and streams originating in Himalayas.

To prevent stream bank erosion, following measures are adopted :

(i) Afforestation and grassland improvement.

(ii) Soil conservation measures in the catchment areas.

(iii) Improving the capability of the channels to carry floods.

Torrents

They are popularly called as chaos in North and North-Eastern India. They carry large flows at high velocity and turbulence. The torrents cause destruction to life and property in valley reaches as a result of serious soil erosion hazard. In India, about 2.73 m. ha of land is affected by the menance of torrents.

This problem is showing upward trend. It is basically due to inherent geological and climatically features and further enhanced by human intervention, both on mountain slopes and foot hills. Torrents could be controlled to a large extent by adopting following remedial measures :

(i) Constructing spurs.

(ii) Constructing marginal bunds at a reasonable distance away from the edge of the banks.

(iii) Providing structures to control channel flow at permissible velocities.

(iv) Improving grasslands.

Four Classes of Land Suitable for cultivation

The US Soil conservation has established a system to classify land use possibility. The land is classified into eight classes and lists the characteristics and capabilities of each.

(i) Land class -I : It is suitable for cultivation. Its main characteristics are that it is excellent, flat and well-drained. It is capable for cropland. The conservation measures required for this class of land is nil.

(ii) Land class - II : This land is suitable for cultivation. It is good land with minor limitations such as slight slope, sandy soil with poor drainage. It is good for crop and pasture. This land is useful if strip cropping and contour farming is done because they are the two conservation measures.

(iii) Land class III - It is moderately good land with important limitations of soil slope or drainage, capable for cropland, pasture and watershed. It needs four conservation measures such as contour farming, strip cropping, terraces and water ways.

(iv) Land class IV - It is fair land with severe limitation of soil, slope, or drainage. This land is capable for pasture, orchards, urban, industry and limited cropland. Conservation measures are contour farming, strip cropping, terraces and waterways.

Four Classes of Land not suitable for cultivation

(i) Land class V - To be used for grain and forestry; limited by rockiness, shallow soil. Has capability for grazing, forestry, watershed, urban and industry. No special precautions of properly grazed or logged; must not be ploughed.

(ii) Land class VI - It has moderately steep slopes. This land is fit for grazing, forestry, watershed, urban and industry. Graining or logging may be done in a limited manner for conservation.

(iii) Land class VII - This class of land has severe limitations for grazing and forestry because of very steep slopes exposed to erosion. This land is capable for grazing, forestry, watershed, recreation wildlife, urban and industry. Careful management is required when used for grazing or logging from soil conservations point of view.

9

FOOD RESOURCES

Food means substance taken to the body to maintain life and growth. Food is necessary to supply energy to both living and non-living organisms. Food is essential to maintain vital functions and to rebuild tissues worn out in the day processes of living. Food intake is measured in calories. Food provide energy as already stated. The amount of energy each of us needs to remain strong and healthy depends on body weight, climate, state of health, stress level and basic metabolism. A small sedentary adult living in a warm climate a large, muscular person living in a vigorous outdoor life in a cold climate might need 6000 to 7000 calories per day to stay warm and active. FAO of UNO estimates that the average minimum daily caloric intake over the whole world is about 2500 calories per day.

Diet

Diet occupy important position and has a great effect on human health. There is a great correlation between cardiovascular disease and animal fat and salt in one's died. Highly processed foods, fat, smoke cured and high nitrate meats are prone to cancer. While grains, complex carbohydrates, fruits and vegetable often have beneficial health effects. Important dietary components are pectins; vitamins A, C and E ; substances extracted form vegetables such as cabbage, broccoli, cauliflower and brussels sprouts and selenium from plants have anticancer effects. Well-to-do people everywhere eat too much. This is often observed in developed countries. At least one-fourth of all Americans are considered overweight. Cutting back on the number of calories consumed reduces strain on bones including cardiovascular arrest, diabetes and may be cancer.

In some areas of the world, people have exceptionally long lives. They are Abkhasian of Soviet Geogia, Hunzans in the mountains of Pakistan and the Vilcabama villagers in Ecuador. Though some of these claim to be 120-140 years old yet, it is difficult to know the dates of their birth. Reasons of their high longevity are given below :

(i) Uncomplicated lifestyle as it reduces conflict and anxiety.

(ii) Enjoy clean air and pure water.

(iii) Diet is simple with low salt and fat: high fiber contents, little meat and lost of vegetables and fruits.

(iv) The whole family shares in decision making, recreation and religion.

(v) Do practical and physical work together for the common benefit of their community.

(vi) Low level of stress.

(vii) Uniform and predictable life in small villages.

(vii) Non-industrialized setting.

(viii) Active and vigorous life with low pressure.

(ix) Living at moderately high elevations with cool, sunny and dry climate.

Observe and experience the following national health recommendations and diet goals.

(i) Eat less radiative food.

(ii) Eat less red meat.

(iii) Eat more fish, poultry, beans and fruits.

(iv) Eat more whole grains, cereals, fruits and vegetables.

(v) Eat less sugar.

(vi) Eat less salt.

(vii) Eat less fat and cholesterol. Also Eat only calories enough to meet the need of body.

Highlights about food

1. Vitamins are organic molecules essential for life. Food sources of vitamins are fruits, milk and green vegetables.

2. Humans also require inorganic nutrients both for building cellular structures and for regulating many cellular reactions.

3. Lipids are an important source of energy. Most of us get adcquate levels of lipids in out diet, but very poor people in some areas of the world exist on diet of starchy food.

4. Most people in the world derive the bulk of daily calorie intake upto 80% from carbohydrates, mainly starches.

5. In addition to calories, we need specific nutrients in our died such as proteins, vitamins and minerals. The average adult human needs 40 g of protein per-day.

6. Indians takes 2100 calories on an average while being in urban cities and 2400 calories in rural areas.

7. The average daily calorie intake in North-America and Europe is above 3500 calories, nearly 1/3 rd more than is needed for adequate nutrition. As least 20% of Americans are seriously overweight.

8. Grain supplies alone could provide everyone in the world with 2700 calories if that grain were equitably distributed.

9. People who receive less than 80 percent of their minimum daily calories requirements are considered seriously undernourished.

10. People who receive less than 90 percent of their minimum dietary intake one a long term basis are considered undernourished.

11. 60-90% of the food that consumers eat, digest and absorb is oxidized for energy, the remaining 10-40%, which is converted to body tissues is very important. This is the fraction that enables the body, to grow, as well as to maintain and repair itself.

Water food resources : By eating balanced diet, we can avoid malnutrition and the ill effects of obesity and cardiovascular diseased.

Food Security

In the past 50 years, the food grain production in the country increased substantially from 51 million tonnes in 1950-51 to 209 million tonnes in 1999-2000. During the same period the population if India increased from 361 million in 1951 to 1027 million inn 2001. Inspite of the constant rise of population, we have been able to build a food stock of 44.7 million tonnes in 2001. The country, on an average has enough in stock to meet the maximum food requirements of its citizens. But this situation is not likely to remains so easy in the years to come. The population of India by 2020 is likely to be around 1300 million. This would require a huge amount of food grain along with pulses, oilseeds, vegetables and fruits.

In spite of surplus food grain stock, it is also a fact that a vats number of people do not have enough money to feed themselves twice a day. Even inn 1999-2000, 26 percent of the population of India was below the poverty line. This ranged between 27 percent inn rural areas and 23.6 percent in urban areas of the country. Though the percentage of people below poverty line declined substantially form 36 percent in 1993-94 to 26 percent in 1999-2000, yet by no means a small number.

There has been a gradual shift from cultivation of food crops to cultivation of fruits, vegetables, oil-seeds and crops, which also act as industrial raw materials, This has led to the reduction in net sown area under cereals, millets and pulses. With the growing population puts a big question mark over the country's future food security. The food production has started to show a declining trend. In 2000-01, India produced only 199 million tonnes of food grain. The use of more and more land for construction of factories, warehouses and shelters has reduced the land under cultivation and new fertile land for farming is no longer available. The productivity of land has started showing a declining trend.

Fertilizers, pesticides and insecticide, which once showed dramatic results, are now being held responsible for degrading soils. Periodic scarcity of water has led to reduction in area under irrigation. Inefficient water management has led to reduction inn area under irrigation.

Under these situations, the key concern for agricultural scientists is to produce more food to satisfy the hunger of more than a billion people, through agricultural techniques which are environmentally sustainable. Biotechnology, as a supplementary tool, can be used for modifying different crops genetically to increase the yield per hectare. This will easily meet out our food, fodder and fibre requirements. Use of such technology has made the corps more resistance to insects, pests and diseases. This will greatly reduce the dependence on insecticides. Genetically modified crops require less water compared to other crops. This, ultimately, will reduce the cost of production and will help to improve the environment. Biotechnology is poised to give benefit to both rich and poor farmers equally, and is more environmentally safe and sustainable.

Hunger and Nutritional deficiency

The nutritional deficiency which means calories intake less than 1000 can effect individuals anywhere inn the world. Reason being that adequate diet is not available due to non-availability of wholesome food. Another reason is that people do not know how to choose right thing to eat. Food and water are our barest necessities to remain alive. One cannot easily imagine what would ti be without it. Let us a have a view about it.

(i) We usually think of hunger in terms of catastrophic famine and starvation, far more people in the world are affected by chronic food shortages in which they lack adequate nutrition on a long'term basis. About 1 million people are thought to have died in Sub-Saharan Africa in the early 1980's as a result of drought related famine.

(ii) Similarly, the world was horrified by terrible photographs of starving children in Ethiopia and Sudan inn 1983 and 1984. More than 1 million people died in the Horn of Africa. Total food supplies dropped by 10 percent, the famine was severe.

(iii) The state of Maharashtra inn central India suffered a similar drought in 1971-73 that reduced crop yields by 50%.

(iv) Adverse weather, insect infestations and other

natural disasters cause crop failures and create food shortages. Problem arise because poor people can neither grow their own food nor find jobs to earn money to buy food they need. In 1974, a terrible famine struck Bangladesh and thousands died. The immediate cause was floods in June though August that interfered with rice planting, cultivating jobs on which many farmers depended. Fears about impending rice shortages triggered panic buying, hoarding by speculators and rapidly rising prices.

(v) Famines are characterized by mass migrations as starving people travel to refugee camps in search of food and medical care. Many die on the way or fall prey to robbers.

(vi) Starving people eat their seed again and slaughter their breeding stock to keep themselves and their breeding stock to keep themselves and their- families alive.

(v) Acute food shortage cause famines, characterized by large-scale loss of life.

(vi) Food shortages are the tragic scenes of mass starvation that come from countries like Ethiopia, Somalia, Sudan, or Mozambique.

Land and Major Crops

We have three crops seasons in India. They are Kharif, rabi and zaid, Khariff season starts with the onset of the monsoon and continuous till the beginning of winter. Cross such as rice, maize, millets, cotton, groundnut, moong urad etc., are grown in this season. Rabi season starts with the beginning of winter and continues till the beginning of Summer. Crops such as wheat, barley, gram and oilseeds are grown in this season. Zaid is short crop season of summer.

Watermelons and cucumbers are grown on the sandy tracts also the river channel. Rice, maize, vegetables and fruits and major crops.

Broadly, there are two types of crops. Kind and cash crops, Kind crops are : Rice, Wheat, Millets, Maize and cash crops are : Sugarcane, Rubber, Tobacco, Spices and Fruits.

Nutrition position In India

Although the country has been able to eliminate nutritional deficiency syndromes like pellagra, beriberi, scurry etc. yet chronic energy deficiency among adults, undernutrition among children and micronutrient deficiencies such as goiter, blindness due to vitamin-A deficiency and anemia are still prevalent across all section of the population. The country has adopted a multi-sectorial and multi- pronged strategy to combat these problems and to improve the nutritional status of the population.

One half of the children under the age of five years in India are moderately or severely malnourished, 30 percent of newborn children are significantly under weight and nearly 60 percent of women are anaemic. Under the Integrated Chile Development Scheme the Department of Women and Child Development implements the food supplementation programme for improving nutritional status of mothers and children. Ministry of Health and Family Welfare implements programmes for tackling major micronutrient deficiencies such as Vitamin A and Iodine deficiency etc.

To address the needs of undernourished infants and children below 3 years. 15 percent of the total Additional Central Assistance are earmarked for the nutrition component under the Pradhan Mantri Gramodaya

Yojana. ACA under the nutrition component of PMGY is an additionally for providing nutrition to children in 0-3 years. States are required to ensure that all below poverty line children by supplementing the States commitment for supplementary nutrition with adequate ACA under PMGY. During 2000-01 and 2001-02 the Department of Women and Child Development was the nodal Department for the implantation of the nutrition component of the programme. From 1st April 2002, the Planning Commission is directly implementing the scheme.

A National Nutrition Mission will soon be launched in selected nutritionally backward districts of the country to further carry forward the goals of National Nutrition Policy. The Planning Commission has allocated Rs. 103.33 crore for 2002-03, as Special Central Assistance to States to provide free foodgrains to undernourished adolescent girls and pregnant & lactating mothers in 51 identified districts. The first installment of Rs. 53.96 lakh has been released to all States for 2002-03.

Energy Conservation

Energy saved is energy produced is a common dictum. Energy conservation means saving of energy because we are short of energy in relation to its demand. Following tips may be adopted to conserve energy :

(i) Buy locally made, long lasting materials.

(ii) Eat locally grown food in season.

(iii) Use of recycled glass, metals, and paper : compost organic wastes.

(iv) Take shorter, cooler showers, install water saving faucets at shower heads.

(v) Stop faucet leaks, especially hot water.

(vi) Turn off light, television sets, and computers when not in use.

(vii) Insulate your house.

(viii) Joining a car pool or driving a smaller, or more energy efficient car.

(ix) Use of stairs instead of elevators if you are not sick.

(x) Use of public transportation, walk, or riding a bicycle.

Role of Individuals in Conservation of Resources and sustainable life style

The land which nature has endowed us is a common heritage. We have shared it with the past generations and will have to do so with the future generation. Human greed has increased so much so that we have almost stopped to think for the future. This is not a healthy trend. In an agricultural country like courts, we should not allow its depletion and degradation. Besides there are many natural resources such as energy, forests, water, soil and biodiversity. If every individual start thinking them as national wealth and a great use of generations to come then these precious resources can be conserved. We now take one by one how these should be conserved.

Home Energy Saving Through Energy Efficiency

Home energy efficiency means designing and building houses that incorporates methods and materials that have been proven to conserve energy. But most of us have to live in house and apartments that already exist. How can they practice home energy conservation. A few practices mentioned below can be adopted.

(i) Emphasize an greater use of non-conventional

sources of energy.

(ii) Switch off electricity whenever not required.

(iii) Use power saving devices.

(iv) Use more and more public transport system and less individual vehicles.

(v) Cogeneration which is as simultaneous production of both electricity and steam from the same plant is one of the fastest growing source of energy.

(v) For greater saving of energy, new houses can be built with extra thick super-insulated walls, air-to-air heat exchangers to warm incoming air, and even double-walled sections that create a house within a house.

(vi) Reducing air infiltration is usually the cheapest, quickest, and most effective way of saving energy because it is the large source of the oil.

(vii) Secondary recovery method may be used to recover a greater amount of the oil.

(viii) Municipal trash can be burned to produce heat and electricity. This will increase the additional supply of energy.

(ix) The easiest way to save energy is to keep an eye on energy-use habits.

(x) High- efficiency fluorescent lights emit the same amount of light but the quite less energy compared to conventional incandescent bulbs.

(xi) Some energy-efficient applications can save substantial amount of energy. Air conditioners and refrigerators with better condensers and heat exchangers use half to one fourth as much energy as older models. New improved furnaces can offer 96 percent efficiency, compared with conventional 70 percent efficiency in older models.

Conservation of water resources

In view of the limited water available, and that too, unevenly and polluted, there is need to adopt certain individual conservation measures because freshwater available on earth today is very limited. The amount of freshwater available on Earth today is no more than that was available 2,000 years ago when the Earth's population was less than 3% of its current size.

Rising demands of water in agriculture domestic consumption, and industry are forcing stiff competition over the distribution of scarce water resources, both amongst different regions and types of use. Water is likely to play the same role in world economy in the 21st century that oil played in the 20th century and will be the source of conflicts world over to pose a major threat to human security if proper preventive measures are not taken globally. Growing water scarcity has become very real.

Rainwater Harvesting

It is a technique of increasing the recharge of ground water by capturing and storing rainwater by constructing structures, such as dugwells, percolation pits, check dams, Rainwater is stored in groundwater reservoirs by adopting artificial recharge techniques to met the household needs through storage in tanks. The objectives of rainwater harvesting are :

(i) To supplement domestic water requirement during summer and long dry spells.

(ii) To improve the quality of groundwater.

(iii) To reduce groundwater pollution.

(iv) To augment the groundwater storage and raise the water table.

(v) To avoid flooding of roads.

(vi) To reduce run off.

(vii) To meet the increasing demand for water.

Watershed development

River basin is an area drained by a river and its tributaries. The watershed is the basin of a tributary. It may have a small stream or it may not have any such stream, but whenever it rains, the water flows through it finally to join some streams. The watershed, thus, is a physiographic unit and can be used conveniently for integrated development of small natural unit areas. The watershed development is holistic approach. It is includes programmes for soil and moisture conservation, water harvesting, afforestation, horticulture, pasture development and upgradation of community land resources. These plans take into consideration the land capability and the local needs of the people. It seeks participation of the local people. Several such programmes have been taken up recently by the central and state governments, with promising results.

Wildlife conservation

India has more than 89,000 species of animals, 1200 species of birds and 2500 species of fish. Besides, there are large varieties of amphibians, reptiles and mammals. Depletion of forests has endangered plant and wildlife. In order to preserve natural habitat and protect them from becoming extinct, the country has developed a network of 89 national parks, 490 sanctuaries and 13 biosphere reserves. The Biosphere reserve programme was launched by the UNESCO in 1971 to conserve representative samples of co-systems, to provide long term in-situ conservation of biodiversity and to promote appropriate and sustainable management of living resources.

In India first biosphere reserve-Nilgiri biosphere reserve came into existence in 1986. So far India have twelve biosphere reserves. Besides this a number of steps described below have been initiated to check wild life conservation.

During the year, ban on export of 29 species of plant, plant-portions and their derivatives obtained from wild, was continued. Export of six species of exotic birds was also continued subject to pre-shipment inspection and provision under CITES wherever required.

The Wild Life Amendment Bill, 2002 was passed by the Parliament during the Winter Session, 2002. The Bill proposes to enhance penalties for violation of the provisions of the Act. It also proposes to create two new categories of Protected Areas, viz., Conservation Reserve and Community Reserve.

A number of species reported as endangered were provided highest protection by including them in the Schedule I and part II of Schedule II of the Wild Life. The Standing Committee of the Indian Board of Wildlife met three times to discuss important aspects relating to wild life. The IBWL Apex Advisory Committee is headed by the Hon'ble Prime Minister. In order to monitor the implementation of the National Wild Life Action Plan, a Monitoring Committee has been set up.

There are 89 National Parks and 500 wildlife sanctuaries in the country covering an area of 1.56 lakhs square km. Financial assistance for development of National Parks and Sanctuaries was provided to 210 National Parks and Sencturies in 25 States including North-Eastern States and Union Territories in the country during the year.

Project Tiger was launch in 1973 on the recommendations of a special task force of Indian Board of wildlife to ensure maintenance of available tiger population in India and to preserve the areas such biological importance as a national heritage for the benefit and educational employment of people.

Conservation of forests

Conservation of forests is a national problem. It must be tackled with perfect coordination between forest department and other departments. People's participation in the conservation of forest is of vital importance. We must celebrate Van Mahotsava with great enthusiasm and zeal and plant trees in all such areas, which are unsuitable for cultivation of crops. We should also create public awareness about the importance of trees. All national day celebration programmes should follow a programme of tree plantation. Progress of forest conservation is detailed below :

(i) Forests Survey of India, Dehradun organized an International Training Workshop on Assessment of Trees Outside Forests from April 22, 2002. The Workshop was attended by 22 participants from 11 countries from South and South East Asia. Food and Agriculture Organisation of the United Nations sponsored the Workshop.

(ii) During the year the inventory data consisting of tree species and both collected from 51 districts, rural and urban areas of all 14 physiographic zones were used to estimate average number of trees per hectare in that zone.

(iii) Special studies for assessment of change in the forest cover of seven Biosphere Reserves, thirteen Tiger Reserves, twelve National Parks and fourteen Wildlife Sanctuaries have also been done.

(iv) Field manuals for forest inventory and TOF urban

have been finalized for data collection.

(v) Wood assessment of national forest cover 125 districts of Trees from 22 states and 15 districts of TOF urban from 13 states spread over 14 physiographic zones were surveyed and data was analyzed.

Assistance to Botanic Gardens

The Ministry has constituted an Expert Group to examine and recommend proposals received under the scheme on Assistance to Botanic Gardens to augment ex-situ conservation of rare endemic plants. One time financial assistance is provided under the scheme. The project for establishment of a National Botanic Garden at NOIDA - Delhi with an outlay of Rs. 3,777,85 lacks has been approved for implementation in the Tenth Five Year Plan.

Medical Plants

Medicinal plants have been identified as one of the thrust areas by the Ministry and different programmes encompassing activities in the area of medicinal plants were continued to carry out conservation of medicinal plants found in the forests and protected areas as well as cultivation of these plants in the degraded forest area.

Medicinal Plants Conservation Areas in the southern States of Kerala, Tamil Nadu and Karnataka and Medicinal Plants Conservation Network established for in-situ conservation earlier continued their activities during the year.

Animal Welfare

The Animal Welfare Division became a part of the Ministry of Environment and Forests in the month of July 2002. Earlier the Division was under Ministry of Statistics and Programme Implementation. The mandate of Animal Welfare Division is to prevent the infliction of unnecessary pain or suffering on animals. Under the Prevention of cruelty to Animals Act, 1960 a new set or rules namely, Animal Birth Control Rule, 2001 was notified on 24th December, 2002.

Under the scheme of Animal Birth Control and Immunication of Stray Dogs, during the year, an amount of Rs. 1.30 crore was released to 49 organisations in various states. The first National Seminar on The States of Animal Houses in India was organized in June, 2002. As an outcome of the Seminar, three sub-committees namely : Committee on Alternatives and Harmonistion with Global Guidelines on Alternatives, Committee on Norms of Rehabilitation of Animals used in experimentation.

An Institute is being set up at Ballabhgarh, District, Faridabad, Haryana for providing various Graduate and post Graduate sources on animal welfare. The Institute would also offer incentive specialized Diplomas inn specific streams.

Equitable use of Resources for Sustainable development

Resources both renewable and non-renewable are scarce in supply to match the demand besides being unevenly distributed. Some countries are using more resources and may have little to use. This is because there is drastic change in life-style. Life style changes may involve such things as switching to a more fuel-efficient car, recycling paper, cans and bottles. Similarly underdeveloped developing countries are adopting new life-styles in an unsustainable manner. Studies reveal that developed countries are more rapidly changing their life style. For instance USA has more aeroplanes compared to many Asian countries.

Poor nations produce only small amounts of

greenhouse gases compared to development nations. Reason being that rich countries continue to lead energy consumptive material life styles. Europe and Japan are responsible for about 2/3 of all potential global warming with them. Mismanagement of natural resources, large- scale deforestation, unplanned discharge of residues wastes, handling of toxic chemicals, indiscriminate construction and expansion of settlement activities pose threat to the environment because equitable use of resources is being hit hard by both rich and poor countries. This attempt has shaked sustainable life-styles. See ozone layer is as wide as the united states and as high Mount-Everest and opens over Antarctica every year. Ozone depletion has been occuring since 1960's but water not grown larger.

Dallas nay have 162 days per year that hot that 100 as it does not. Ecological disruptions could be severe. There shall be a rise in sea level due to global warming. A few oceangraphers are of the opinion that thermal expansion alone could raise the sea level by one meter or more in the next century. Global warming might cause increase in violent storms such as Hurricane.

May scientists state that global change in environment has already started. They point to the fact that four of the five warmest years on record were in the late 1980s and early 1990s. A new approach to gauge earth's climate has found that the atmosphere would witness more warming during 21st century. It is said that the global warming would be faster than what has been anticipated till now.

The approach, developed as the Handley Centre for Climate Prediction and Research inn Bracknell in UK and published in Nature, said average global temperature could be 5.5 degree Celsius higher by 2100, which is around 1.5 degree Celsius higher than a forecast accepted worldwide.

Contrary to earlier estimates which are based on limited factors only, the new approach says that the warming of earth's atmosphere would depend on a host of other factors like volcanoes, fluctuations in the sun's activity and changing levels of greenhouse gases and ozone. It also allows oceans to affect the land and vice versa.

James Baker, former administrator of the US National Oceanic and Atmospheric Administration while being in India some time back, commented that the world will see a definite impact of global warming within the next 10 years. Increase in global warming will make society more vulnerable to climate change. Policy- makers all over the world will either have to learn to do with lesser greenhouse gases or learn to live climate change. More scientists are convinced that we are seeing human influence on the climate what we are worried about is the enhanced Greenhouse Effect, With carbon dioxide levels in the atmosphere growing. Barker admitted that Developed countries have no right to tell developing countries to slow down their economic growth.

Developed countries need to find solutions and help developing countries leap frog over polluting technologies in their quest for raising standards of living. Indeed, we are all involved whether we recognized it or, not in degrading environment. Simply by out existence on the planet; everything we do- the car we drive, the product we use, the wastes we throw-away, virtually every choice we make and action we take-has certain environmental impact and certain consequences for the future. Therefore, it is not a matter of having an effect ; but of what and how great that effect will be. It is a matter of each of us asking ourselves.

Life Cycle Assessment

Life Cycle Assessment is a decision cum management too which provides information on the environmental effects of various products and processes so as to arrive at necessary corrective measures to make the entire process efficient with optimal utilization of resources and minimal wastes generation. LCA studies have been initiated in various sectors namely : Steel, Pulp and Thermal Power. The study in the Steel Sector was completed earlier and the report is under preparation. The study relating to the Thermal Power Sector has been completed during the year while the study in the Pulp and Paper Sector is progressing as per schedule.

Development of Natural Dyes from Forest Wastes

A three years duration project has been sponsored to Forest Research Institute, Dehradun for development of natural dyes from forest waste.

Bio-remediation of Raladevi Lake in Thane, Maharashtra

A project relating to cleaning of Raladevi Lake in Thane District in Maharashtra using bio-remediation technique has been sponsored to Thane Municipal Corporation.

Environmental Statement

Environmental audit is a management tool and provides a structure and comprehensive mechanism for ensuring that the activities and products of an enterprise do no cause unacceptable effects on the environment. Submission of the Environmental Statement by polluting units seeking consent either under the Water Act, 1981 or both and the Authorization under the Hazardous

Wastes and Handling Rules, 1989 has been made mandatory through a Gazette Notification of April , 1993 under the Environment protection Act, 1986.

The Environmental Statement enables the units to undertake a comprehensive look at their industrial operations and facilities, understanding of material flows and focus on areas where waste reduction and consequently saving in-put cost if possible. The primary benefit of environmental audit is that it ensures cost effective compliance of laws, standards, regulations, company policies etc.

During the year, action has been initiated to evolve model environmental statements in various sectors for facilitating comparison in use of raw material, water conservation, energy consumption etc. A project has been sponsored to Central Pollution Control Board for evolving model environmental statement in the eight sectors viz. Sugar, Thermal Power Projects, Cement, Paper and Pulp, Pesticides, Bulk Drugs, Tanneries and Textiles Activities.

Recycling of Marble Slurry in Udaipur, Rajasthan

A two years duration project has been sponsored for the manufactures of bricks and tiles from marble slurry in Udaipur, Rajasthan with the aim of utilizing wastes arising out of marble cutting and processing for the purpose of improving the local environment.

Field Demonstration and Development of Bamboo Based Composites/Panels

The ongoing project on field demonstration and development of bamboo based composites/ panels was continued during the year. Under this project, commercial production of horizontal and vertical laminates have been made using Bambusa bamboo

Species. Construction of demonstration house will be taken up during the current year.

Development of Market Based Instruments for Regional Environmental Management in the Kawas-Hazira Region in Gujarat

The ongoing project on Development of Market Based Instruments for Kawas-Hazira Region of Gujarat has been completed. The study report has observed that the taxes and incentives based on efficiency improvements align the pollution control agencies better with the polluters that the Command and Control regime. Such an instrument also facilitates prescribing incentives for achieving the triple bottom line, viz economic-efficiency, environment-responsibility, and social-relevance entitling the Corporate to Clean Development Mechanism and other cleaner-production benefits. The recommendations of the study have been discussed in a workshop with all concerned.

Industrial Ecology Opportunities in Ankleshwar and Nandesari Industrial Estates, Gujarat

A study was undertaken to develop and implement Industrial Ecology Opportunities in Ankleshwar and Nandesari Industrial Estates of Gujarat. The study has examined technologies used by industries in this region and suggested possible approach to achieve eco-efficiency within the industrial estates to reuse and recycle wastes and affluents generated from different industrial units. The recommendations of the study have been discussed in a workshop with all stakeholders including State Government Agencies for their implementation and an Action Plan is being drawn-up by them.

Environmental Statistics and Mapping

For sound Environmental Management, reliable information base and the mapping of areas needing special attention for pollution prevention and control are a prerequisite. As a step in the direction, projects and pilot studies have been initiated various research institutions and organizations. Under this program, following studies have been initiated and are in various stages of completion :

GIS base Hydrological Modeling for Water Quality and Quality inn Cauvery River Basin by IIT, Delhi. Geochemical baseline Mapping for Environmental Management by National Geophysical Research Institute, Hyderabad.

Waste Minimization/Cleaner Production

Waste minimization is one of the strategies adopted for minimizing the industrial pollution. The objective of the scheme is to assist the small and medium scale industry inn adoption of cleaner production practices. A project has been sponsored to National Productivity Council on Waste Minimizations in Small Scale Industries for establishment and running of waste minimization circles in clusters of small scale industries, capacity building in area of cleaner production, establishment of demonstration units in selected industrial sectors etc.

So far 115 waste minimization circles have been established through out the country and a large number of Organizations and Institutions have been trained in waste minimization activities. The project was executed till November, 2002 under the World Band project on Industrial Pollution Prevention as Phase - I and Phase - II is now being continued with Ministry's internal funds. Implementation of this project has helped in identification of more than 200 options for resources and energy conservation in various small scale industries.

Industrial Pollution Abatement through preventive strategies

This scheme is an amalgamation of the three on-going schemes viz, Environmental Audit, Adoption of Clean Technologies in Small Scale Industries and Environmental Statistics and Mapping, which have been continuing since eighth Five Year Plan. Due to encouraging results and benefits to various small scale units, these schemes are being continued during the 10th Five year Plan also.

Goals toward Sustainable life-styles and development

(i) Take a leadership role in the development and implementation of global sustainable development policies.

(ii) Move towards stabilization of population.

(iii) Create full opportunities for citizens, business and communities to participate in and influence the natural resources, environmental and economic decision that effect them.

(iv) Encourage people to work together to create healthy communities where natural and historic resources are preserved, education is life long, transportation and health care are assessable, and all citizens have opportunities to improve the quality of their lives.

(v) Use, conserve, protect and restore natural resources in ways that help ensure long-term social, economic, and environmental benefits for ourselves and future generations.

(vi) Sustain a healthy economy that grows, sufficiently to create meaningful jobs, reduce poverty and provide the opportunity of high quality of life.

(vii) Ensure that every person enjoys the benefit of clean-air, clean-water and healthy environment at home, at work, and at play.

10

ENVIRONMENT AND HEALTH CONCEPTS

Introduction

Today various kinds of changes have taken place in different spheres of life. With such kinds of changes, dramatic kind of change has taken place in the outlook and thinking pattern of common men. Earlier he was only concerned with finding out the means by which he can live a comfortable life, but now he tries to find out those means by which he live a comfortable and healthy life. People have now become health conscious. One of the main reason of this happening can be the fact that a paramount increase in the number of people suffering from various diseases has been noticed in the recent past. Not only this, as the literary level is increasing in the nation, various kinds of information relating to healthful living is being made available to them, because of which they have understood the value and importance of living a healthy life.

Now even a commonman knows the importance of health. Thus, if you are also concerned about your health, then you are not alone as at this time, majority of people are also concerned with their health. This can be considered as one of the main reasons that why number of private organisations and clubs making arrangements of various physical activities have increased in today's time. People belonging to higher strata of society do not get indulged in such organisations, but people belonging to middle classes or strata of society have also now become highly health conscious. Now everyone knows what to eat, when to eat and how much to eat. Before taking or consuming any food article, everyone give consideration to the nutrients value it possess and also the effects which the food item will mark on the health.

Although number of people becoming health conscious or making use of health services has increased to considerable extent, but still number of people suffering from problem of obesity and various other diseases arising out of this problem has increased to considerable extent, because of which maintaining health and leading a healthful life has now become a challenge for us. Various people will pose and ask the question that why is being healthy so challenge for so many of us. What can we do to overcome these challenges, make better decisions about our behaviour. There are no easy answers, because health is influenced by various factors, out of which some that we can control, some that we cannot. But the good news is that many people have found the skills and motivation to look objectively at where they are, plan carefully, and make decisions that improve their life and health.

As health has become so important concept in today's life, because of which it has attracted the attention of various researchers and experts. It is on the basis of best available research that health decisions should be based. All the decisions taken on such bases should be consistent with who you are, your values and beliefs, and who you want to become. Although health is not always totally within your control, some behaviour choices that affect health are certainly under control. The choices one makes can enhance his health, as well as reduce future health threatening risks. For those risk factors that are beyond one's control, one can learn to react, adapt, respond in appropriate manner and use a reasoned, rather than purely emotional, rationale method for taking the decisions. Quality and length of the life can be improved with the help of making informed and rational kinds of decisions. Not only this, such decisions also effect the persons around you in a positive manner.

As is clear from the preceding chapters that although term health is used by almost all the people of world, but there are very few out of them who understand properly what it really means. For some, health simply means the antithesis of sickness. To others, it means being in good physical shape and having an ability to resist illness. Still others use term such as wellness or well being to include a wide array of factors that seem to lead to positive health status. The main reason of existence of such kind of differences lies in the fact increasingly enlightened way of viewing health has taken shape over time. In addition, as collective understanding of the concept of illness has improved, so has human being's ability to understand the many nuances of health. In reality, one's progress to current understanding about health has evolved over centuries, and a long way has to be covered even till now in order to achieve a truly comprehensive view of this complex subject. Much discussion on the topic of various concepts relating to health had been discussed in the preceding chapters, because of which, they will not be discussed at length in this chapter. This chapter is designed to provide fundamental knowledge about various health topics, so that you can utilise personal and community resources for creating your own health profile, and to challenge you to think more carefully prior to making any kinds of decisions that affect your health status.

Evolution of Health Concept

Prior to 1800s, a person who was not sick was being regarded as not only lucky, but also healthy. When deadly epidemics such as plague, influenza etc., killed millions of people, survivors were believed to be of hearty, healthy stock, and they congratulated themselves on their good fortune. Poor health was not often associated with poor hygiene and unsanitary conditions, and certain stigmas were attached to households that harboured any of these illnesses. It was not until the late 1800s did researchers discover that victims of these epidemics were not simply unhealthy or dirty, rather they were victims of environmental factors which made them sick and over which they often had little or no control. Public health officials moved swiftly to address these problems, and as a result, the term Health became synonymous with good hygiene. Various schools began of offer courses in health and hygiene. It was into the 1900s that investigation into the environment as the primary cause of disease continued to take place, which was as a result of outbreak of tuberculosis and influenza surged i many regions of the world. People who made it through the first few years of life without succumbing to an infectious disease usually were able to survive to old age. Continued improvements in sanitation brought dramatic changes in life expectancy, and the development of vaccines and antibiotics added even more years to the average life of common men.

It was in around 1940s that progressive thinkers in public health began to note that there was more to health than hygiene or disease. They took into notice the definition provided by World Health Organisation in one of the conferences held on international level, in which he defined the term Health as the state of complete physical, mental and social well being. Scientists took a long time in accepting the fact that health included much more than physical, social or mental elements of life. Other elements, such as environmental, spiritual, emotional and intellectual aspects, helped to define whether a person was truly capable of optimal functioning. In addition, the new voices and views suggested that it is not just length of life or the number of disease free years that matters, but rather achieving your potential for a happy, healthy and productive life also.

Today, there is treatment of various diseases, which were once considered to be incurable in nature. As a result of such developments, mortality rate has lower down to greater extent. Morbidity rates also indicate that people less frequently contract the common infectious diseases that devastated previous generations, but still it cannot be said that these facts are indications of healthful living people. If one view the conditions in which major portion of population in our nation live, then all these factors, who are regarded as indicators of health, will be banished by the expert. Although number of persons getting dead as a result of various diseases has decreased and most childhood diseases are preventable or curable and massive public health efforts are aimed at reducing the spread of infectious diseases, but still longer life and less frequent diseases are not proof that people are indeed healthy and leading a healthful life.

In accordance with some experts, Health is a quality of life, involving social, emotional, mental, spiritual and biological fitness on the part of the individual, which results from adaptations to the environment. The concept of adaptability, or the ability to successfully cope with ups and downs of the life, became a key element of overall health definition. Eventually, the term wellness became popular which implied that there were levels of health in each category. In order to achieve a high level wellness, a person moved progressively higher on a continuum of positive health indicators. Those who failed to achieve these levels might move to the illness side of the continuum. Today, health are wellness are often used interchangeably to mean the dynamic, ever changing process of trying to achieve one's potential in each of several interrelated dimensions. Physical Health, Social Health, Mental Health, Emotional Health, Spiritual Health are some of the important dimensions of the concept of health. They are not discussed here at length as mention of these has been done separately in the preceding chapters.

Various scientists and experts have put forward the features which generally exist in healthy persons, mention of which are as follows :—

- A healthy person possess a realistic sense of self, including personal capabilities and limitations.

- Such a person possess the willingness to understand imperfections in himself and also in the persons living around him. Not only this, he possess the ability to analyse all the situations.

- Such a person possess the ability to laugh, to cry and to genuinely feel emotion without getting lost in emotional upsets.

- Such a person possess a sense of satisfaction with life and from whatever functions he perform. Not only this, he possess the ability to appreciate stages of life experiences.

- A healthy person has feeling of self-respect as well as respect for others.

- Such a person possess a zest for living a comfortable and successful life, which is being coupled with a curiosity about what each new encounter and each new day will bring.

- Such a person knows how to keep everything in life under control and in balanced position.

- Such a person likes to accept the challenges imposed by the life on him and never tries to escape from harsh realities of the life.

There are number of people who are of the view that wellness can best be achieved by adopting a holistic approach, which emphasises the integration of and balance among mind, body and spirit. Achieving wellness implies attaining the optimum level of wellness for a give person's unique set of limitations and strengths. A physically disabled person may function at his or her optimum level of performance, can enjoy satisfying interpersonal relationships, can maintain emotional, spiritual and intellectual health and have a strong interest in environmental concerns. On the other hand, persons who spend hours lifting weights to perfect the size and shape of each muscle, but pay little attention to nutrition may look healthy, but they may not maintain a good balance on all areas of health. Although generally we consider physical attractiveness and other external trappings in measuring the overall health of a person, appearance and physical performance are actually only two signs of wellness, indicating little about the other dimensions of health in any way.

Promoting Health

The term health promotion or promoting health describes the educational, organisational, procedural, environmental, social and financial supports that help individuals and groups reduce negative health behaviours and promote positive changes in the behaviours. Health promotion programs identify the healthy people who arc engaging in risk behaviours, or behaviours that increase susceptibility to negative health outcomes, and motivate them to change their actions or behavioural patterns.

Effective health promotion programs provide information about risk behaviours and possible consequences of various harmful kinds of behaviours. Such programs also encourage those with sound health habits to maintain them. By attempting to modify behaviours, increase skills, change attitude, increase knowledge, influence values and improve health decision making, health promotion goes well beyond the simple information campaign. By basing programs and services in communities, organisations, schools and other places where people spend their time, health promotion increases the likelihood of long term success on the road to health and wellness.

Whether we use the term health or wellness, we are talking about a person's overall response to the challenges of living. The ability to recognise that each of us is an imperfect being, attempting to adapt in an imperfect world, signals individual well being. We must also remember to be tolerant of others. Rather than be warriors against pleasure in our zeal to change the health behaviours of others, we need to be supportive, understanding and non-judgemental. Health basing, i.e., intolerance or negative feelings, words or actions aimed at people who fail to meet expectations of health, may indicate our own deficiencies in the psychological, social and spiritual dimensions of the health continuum.

Preventing the Diseases

Most health promotion initiatives include the term disease prevention. Historically, the health literature describes three types of prevention, namely, primary, secondary and tertiary. In a general sense, prevention implies taking positive actions now to avoid becoming sick later. Getting immunized against diseases as polio and chickenpox constitute primary prevention, which are in reality actions designed to reduce risk and avoid health problems prior to they begin. Secondary prevention involves recognising health risks or early problems and taking action to stop or to bring them to an end on the behaviour it leads to actual illness. The third kind is tertiary prevention, which involves treatment and rehabilitation after the person is already sick, which is generally offered by medical specialists.

Number of deaths taking place as a result of wrong habits of consuming harmful substances is increasing in recent times in our nation, which implies that primary and secondary prevention offer our best hope for reducing such incidences. Time has come when we need to move from a mind set of tertiary prevention to focus on earlier intervention designed to remove barriers and help individuals succeeded in their behaviour change strategies.

Health Behaviours

Individual behaviour is a major determinant of good health, but heredity, access to health care and environmental factors can also influence health status. When these factors are considered together and form the basis of a person's lifestyle choices, the net effect on health can be great. Most experts believe that several key behaviours will help people live longer, some of which are as follows :—

i. Getting a good night's sleep, which should be minimum of seven hours.

ii. Maintaining health eating habits, by which human being can attain all essential nutrients by which he can develop at proper pace and manner.

iii. Participating in physical recreational activities, for which membership in the club or organisation established for this purpose can be taken.

iv. Avoid substances having injurious and harmful effects on the health.

v. Limiting intake of alcohol and to practice safer sex by making use of various intraceptives.

vi. Forming and maintaining meaningful and healthful relationships with family and friends.

vii. Controlling the real and imaginary stressors in life.

viii. Participating in atleast one fun activity every day.

ix. Enjoying the environment in which you live and also respect the others around you.

x. Understand the health care system and to use it wisely.

xi. Be as kind to yourself as to others.

Although it is much more easy to list things that one should do, and it is not very easy to bring changes in the behavioural pattern. All of us, irrespective of the fact that where we are on the health continuum, have to start from somewhere. All people have faced personal and external challenges to their attempts to change health behaviours. Some have not done so well, some have been extremely successful, and some have made only small changes that add up to significant improvements in how they feel and how they live. But still, it is very necessary to adopt some of the healthy habits in life, for which behaviour should be modified in a positive and encouraging manner.

How to Prepare for Changing Behaviour

The possibilities of successfully changing negative behaviour improve when you make gradual changes that give you time to unlearn negative patterns and to substitute positive ones. In order to understand the process of behaviour change, it is first important to identify behaviour patterns and attempt to understand the reasons for them. There are various factors which affect or influence behaviour change to considerable extent. Experts have divided these factors into three categories, namely, Enable Factors, Predisposing Factors and Reinforcing Factors.

Enabling Factors

All the factors which make a behaviour change decision more convenient or difficult are included in this category. Enabling factors include one's physical, emotional and mental capabilities, one's skills and one's access to resources. Positive enablers encourage a person to carry through on his intentions, while negative enablers work against one's intentions to change. One must possess the ability to identify the positive and negative enabling factors and to devise alternative pans when the negative factors outweigh and the positive factors are part of planning for behaviour change.

Predisposing Factors

One's life experiences, knowledge, cultural and ethnic heritage and current beliefs and values are all included in the category of Predisposing Factors. These factors influence behaviour and behavioural changes to considerable extent. Age, sex, race, income, family background and educational qualifications are some of the important factors which can play major role in predisposing a human being.

Reinforcing Factors

Presence or absence of support, encouragement or discouragement from significant people in your life are all some of the important reinforcing factors. The manner in which one reward or punish oneself also plays an important role. Accepting small failures and concentrating on the success can foster further achievements. Telling yourself that you are worth the extra time and efforts and giving yourself a pat on the back for small accomplishments are often overlooked factors in positive behaviour change.

It is very necessary to mention that want and an urge to change is a prerequisite of the changing process, but there is much more to the process that motivation. Motivation must be combined with common sense, commitment, and a realistic understanding of how best to move to healthy lifestyle. Readiness is the state of being that precedes behaviour change. People who are ready to change possess the knowledge, attitudes, skills and internal and external resources that make change a likely reality. To be ready for change, a person must take certain basic steps and make adjustments in thinking pattern.

Generally it is assumed that when rational people realise what they are doing carries a risk, they will act to reduce that risk, but this is not necessarily true. There are number of health professionals who even smoke and consume high fat diets. They surely know better, but their knowing is disconnected from their doing. There are two strong influences on behaviour which are found to play effective role by the experts. These two influences are beliefs and attitudes. A belief is an appraisal of the relationship existing in between some object, action or idea and some attribute of that object, action or idea. On the other hand, an attitude is a relatively stable set of beliefs, feelings and behavioural tendencies in relation to something or someone. Various studies have depict the relationship between beliefs and health habits, which show that although beliefs can subtly influence behaviour, they may not actually cause people to behave differently.

From a study, it has been concluded that several

factors must support a belief before change is likely, which include the perceived seriousness of the health problem, perceived susceptibility to the health problem and cues to action. Three other factors are linked to perceive risk for health problems, which include demographic variables involving qualities like age, gender, and race; socio-psychological variables, involving features like personality traits and social pressure and structural variables involving knowledge about or prior contact with the various health problems.

A person's attitudes reflect his emotional responses to situations and follow from his beliefs. According to an expert's views, our behaviours result from our intentions to perform actions. An intention is a product of our attitude toward an action and our beliefs about what others may want us to do. A behavioural intention, then, is a written or stated commitment to perform an action. The more consistent and powerful your attitudes about an action and the more you will be influenced by others to take that action, the greater will be your stated intention to do so. The more you verbalise your commitment to change, the more likely you are to get success.

There are various people who are highly influenced by the approval or disapproval of close friends, loved ones and the social and cultural groups to which they belong. Such influences can support healthy behaviour or they can interfere with even the best intentions. From the time of birth, a child's parents give him strong cues about which actions are socially acceptable and which are not. Brushing the teeth, bathing, wearing clothes and chewing food with your mouth closed are all behaviours that a child learns from the family members. A person's family culture influences his food choices, religious beliefs, political beliefs and all other values and actions. If person gets deviated from his family's norms, then any member of the family will probably let him know fairly quickly. Good family units share a dedication to the healthful development of all family members, unconditional trust and a commitment to work out difficulties in the life. When the loving family unit does not exist, when it does not provide for basic human needs, or when dysfunctional, irresponsible individuals try to build a family under the influence of drugs or alcohol, it becomes difficult for a child to learn positive health behaviours in such environment. Healthy behaviours get their origin in healthy homes, unhealthy homes breed unhealthy habits. Healthy families provide the foundation for a clear and necessary understanding of what is right and what is wrong, what is positive and what is negative. Without this fundamental grounding, many young people have great difficulties in adopting healthy behaviours.

A child's world remains limited to the house or to the family members but when he gets entered into the stage of adolescent, then his world boundaries get extended. He comes into direct contact with other members of the society. Thus the conditions and atmosphere in which he now stays got distinguished from the environment prevailing in his family or at house. In such kind of situation, he has to maintain healthy and co-operative kinds of relationships with the other members. Mostly at this stage, he encounters the persons with varying interests and views. It has been found that more a person values the opinions of other people, the more likely he will be to change a behaviour that offends them. If a person cannot care less what others think about him, he probably brush off their negative reactions or suggested change. Often people say that they do not care about the others, but in reality, most of us care too much about what even the insignificant people in our lives think. In simpler terms, it can be said that lower your self esteem and self efficacy, the higher the chances that others will influence your actions. Sometimes, the influence of others can be a powerful social support for positive behaviour changes. At other times, we are influenced in some wrong practices because we do not want to be left out or criticised. Learning to understand the subtle and not so subtle ways in which other people influence our actions is an important step toward changing behavioural pattern.

Important Techniques for Changing Behaviour

After analysing all the factors that influence what you do, you must decide which behaviour change technique will work best for you. There are some specific kinds of techniques or strategies which help in bringing various kinds of changes in the behavioural pattern of human beings, mention of some of which are as follows:—

i. Visualisation : Mental practice and rehearsal can help in changing unhealthy behaviours into healthy ones. Athletes and others use a technique known as Imagined Rehearsal to reach their goals. By visualising their planned action ahead of time, they are better prepared when they put themselves to the test. For instance, if someone has to appear in an interview then he will make all efforts by which he can improve his manner of communication. For this purpose, he will make rehearsal the manner in which he will reply the questions of concerned person, in which he can take the help of his friend also. Thus, through this technique, he will be able to reply the answers imposed in the interview more effectively and efficiently.

ii. Shaping : There are some behaviours to which it is impossible to change immediately. To reach your goal, you may need to take a number of individual steps, each designed to change one small piece of the larger behaviour. This process is known as Shaping. For instance, suppose that you have not exercised for a while. Now you decide that you want to get into shape, and your goal is to walk three to four miles every day. You realise that you have face a near death experience if you tried to walk even just a few blocks in your current condition. So you decide to start slowly and build up to your desired fitness every other day at a slow and relaxed pace. After some time, you will walk for the same amount of time, but will speed up your pace and cover slightly more ground. After some more time, you will speed up even and will try to go even further. In this way, you will continue taking such steps until you reach your goal. There can be difference in the desired behaviour change, but all shaping involves certain specific kinds of items, mention of which are as follows :—

- Beginning with slow pace and trying not to cause undue stress during the early stages of the program.

- Keeping the steps small and achievable.

- Being flexible and ready to bring any kind of change in the plan if original plan proves to be uncomfortable in any way.

- Refusing to skip steps or to move to the next step until the previous step has been mastered.

It must be kept in mind that as behaviours do not develop overnight, so it is not possible to change them also overnight.

iii. Reinforcement : A reward given with the purpose to increase the likelihood that a behaviour change will occur is known as a Positive Reinforcement. Each of us is motivated by different reinforcers. Experts have classified most positive reinforcers into certain specific categories, which are : manipulative, activity, possessional and social. While making choice of reinforcers, determine the fact that what would motivate you to act in a particular way. Various researches have shown that people can be motivated to change their behaviours if they set themselves up on a token economy system, whereby they earn points that can be exchanged for meaningful rewards, such as financial incentives.

It is quite a difficult task to determine which incentive will be most effective. Your reinforcers may initially come from others, but as you see positive changes in yourself, you will begin to reward and reinforce yourself. Always keep in mind that reinforcers should immediately follow a behaviour, but beware of overkill.

iv. Controlling the Situation : Sometimes, the right setting or the right group of people will positively influence your behaviours. Many situations and occasions trigger certain actions. Usually attempts are made to influence a behaviour by making use of situations and occasions to control it, which is generally known as Situational Inducement. You should know the art of carefully considering which setting will help and which will hurt your effort to change, and by making a firm decision to seek the first and avoid the second, you will improve your chances for chance.

v. Modelling : This is considered to be one of the most effective techniques for bringing positive kinds of changes in the behavioural patterns. This involves learning behaviours through careful observation of other people. Generally people becomes motivated to participate in various health care organisations when they find that someone they know have also joined such any organisation. Thus, this technique can play an effective role in bringing positive kind of changes in the healthful behaviours of people.

Steps for Changing Behaviour

As said earlier that changing behaviour is not a simple process in any way and it comprises of certain steps which should be taken carefully, mention of which are as follows :—

a. Self Assessment : An important fact regarding behaviours, thoughts and feelings is that they always occur in a context of the situation. Experts have divided various situations into certain categories, which are the events that come before and after. Antecedents are the setting events for a behaviour, they cue or stimulate a person to act in certain ways. Antecedents can be physical events, thoughts, emotions, or the actions of other people. Results of behaviour, which are called consequences, affect whether a person will repeat that action or not. Physical events, thoughts, emotions or the actions of other people are included in the category known as consequences. One of an important method of changing behaviour is learning to recognise the antecedents of a behaviour and acting to modify them.

b. Analysing Personal Behaviour : Successful behaviour change requires determining what you want to change. Determining the specific behaviour you would like to change in contrast to the general problem will allow you to set clear and realistic goals.

c. Decision Making : After proper analysation, function of decision making should be performed. Now it is a time to make a decision that will lead to positive health outcomes. It is not easy to make choice among various available alternatives. Just saying no is usually easier and said than done, however, if you are trying to fit in, be liked, or satisfy other emotional needs, decision making will become even more difficult. That is why anticipating what might occur in a given setting and thinking through all possible safe alternatives is important as you implement behaviour change. It should be remember that things typically do not just happen. By staying alert to potential problems, being aware of your alternatives, maintaining a good sense of your own values, and sticking to your beliefs under pressure, you can gain control over many situations in your life.

d. Set Achievable Goals : It is not an easy task to bring any kind of change in the behavioural pattern in any way, but sometimes we make it even harder by setting unrealistic and unachievable goals. Before setting goals, one should analyse the fact that what kind of changes he wants to bring in himself, and which change is the greatest priority at this time in his life. He should also analyse the fact that why he regards one change more important than the other and what will going to be positive outcomes of potential change.

It is only through taking all the above mentioned steps that positive kinds of changes can be brought in the healthy behaviours. People can learn to know how to develop positive and healthy kinds of behaviour only by bringing positive kinds of changes in their outlook, for which they require proper motivation and encouragement. Although health educator is the best person to perform this function properly, but family members and other people around the person can also help in performing this function to considerable and maximum possible extent. They should keep a check on the habits of person in order to ensure that he does get indulged in any kind of harmful behaviour. Not only this, they should pin point the kind of deformities or problems arising in his physical health, as it will encourage the person to make use of some health program or to get participate in some physical activities carried out in organisation or club situated near to his house or place of work.

11

TRANSPORT SYSTEM AND ENVIRONMENT

Essential Elements of Transport

The four elements of transport are (i) the way, (ii) the unit of carriage, (iii) the motive power unit, and the terminal.

(i) The way. Natural ways are cheap and free, and have no maintenance costs unless we try to improvc them artificially. The sea, the air, the rivers, and footpaths are all natural ways. Being natural they are subject to the whims of nature, and this often requires, that they be improved artificially. Rivers are subject to controls to prevent flooding in wet periods and insufficient flow in dry periods.

They are dredged to maintain a channel and locks are built to improve navigation in the upper reaches, Bridlepaths are made up and turned into roads. Highways and motorways, canals, railways, tramways tunnels, and monorails are similarly constructed. Clearly these are not 'free' like may be borne socially rather than privately. If the costs are borne by the ratepayer and taxpayer we may have what is an apparently free way built the owner usually has sole use of it. This owner then charges for its use by other persons, to recoup the capital costs.

(ii) The unit of carriage. Whatever we call it, some vehicle or craft must be used in transport. The efficiency of the mode of transport depends to some extent on the flexibility and adaptability of the unit of carriage used. Road vehicles are adaptable than railways rolling stock because they are not entirely tied to the way : aircraft and ships are even less tightly bound by the way on which they travel.

Even a pipeline can be considered as unit of carriage. It is not very adaptable in its behaviour: we cannot expect it to carry gas until noon, milk from noon till 3 p.m., and petrol from 3 p.m. to midnight. In choosing our method of transport the adaptability of the unit of carriage will be a major consideration.

(iii) The motive power unit. Every vehicle must be driven and the choice of a propulsion unit depends upon the strength of the vehicle, the speed required, the available fuel, and other factors. Today the steam engine, the first great prime mover, has been largely replaced by the petrol engine, the jet engine, the diesel engine and the electric motor.

(iv) The terminal. Nearly every journey involves junctions where we can transfer from one form of transport to another. A port is usually regarded as a terminal for ships, but in fact it is also a terminal for trains, roads, pipelines, and aircraft.

In planning efficient transport systems, commercial firms and transport authorities must view the interchange of facilities as being part of a unified whole. Congestion in terminals in the past has spelt the death of transport system, as it did when the congestion on the canals led to the growth of railways.

The Position of Transport in the Pattern of Commerce

If specialized production is how advanced societies create the utilities they need to satisfy their 'wants,' some way must be devised to bring to the market-place the surpluses created by each specialist. The function of transport is to more goods and passengers geographically.

It is part of tertiary production, satisfying 'wants' by making goods available and by giving service to those who need, for family, business, or recreational reasons, to travel in their day-to-day affairs.

It also enables a fuller use to be made of the division of labour, because transport increases the size of a market and therefore permits a higher degree of specialization to employed. An improved transport system makes possible an increase in the scale of production so that the wealth of the whole nation increases.

Road Transport

Road transport has great advantages over inland raid and waterways, now that petrol, diesel, and jet engines have furnished a variety of cheap motive powers. The chief advantages are as follow:-

(i) Road vehicles are extremely flexible, and computerized route-charts can now be prepared which give the most economic route.

(ii) Specialized vehicles of great variety have been produced by the road-vehicle manufacturers. They include specialized vehicles for containers, packaged timber and other loads.

(iii) Intermediate terminal handling is laminated by door-to-door delivery. The operators' terminal is under his personal direction and control, designed to do what he requires. Delivery is made direct to the consignee's premises, and if these premises are not equipped to handle the goods being delivered this can often be overcome by appliances actually on the vehicle.

(iv) Although slower than rail fright when in actual transit, the slow speed is more than made up by the elimination of transhipment, the directness of the journey and the personal control of the entire movement. Two-way radios even enable drivers to report their exact positions so they can be called to assist other drivers to report their exact positions so they can be called to assist other drivers to pick up new loads.

(v) The 'way' is at present still great, if one ignores vehicle taxation. Proposals that electronic devices should be fitted to vehicles to charge them for use of busy roads have not yet ben implemented.

(vi) Low capital requirements are a feature of the industry, which permits a new entrant to start with a single vehicle. True, he has to face a licensing problem with certain kinds of vehicle, but many new operators do secure licences especially since the industry is continually expanding.

Solving the Inland Transport Problem—Integrated or Co-ordinated Services?

Since the Transport Act of 1947 first se up a nationalized system, inland transport in Britain has been bedevilled by politics. For many years we were in the position of operating a system which had now become law, but was about to be modified by the next

Government.

The discussion revolved around whether we should have an integrated transport system or a number of co-ordinated services each of which was a separate organization. An integrated system of transport would be operated by a central authority with the power to say who should operate certain services.

It would clearly be an enormous organization, and experience has already shown that a system of this size suffers from the diseconomies of large-scale operation.

On the other hand, it does provide an overall body to control the opposing faction in transport systems, much of it socially provided, are wisely spent. An integrated service will be governed by the basic precept that what is needed is a cheap, efficient transport system which provides consumers with goods and services at the minimum personal and social cost.

Three White Papers issued in late 1967 outlined the Labour Government's proposal, which were put into effect by the 1968 Transport Act. This Act proved to be a very effective Act and for several years the turmoil in transport subsided with a reasonable amount of agreement on all sides. The reasons seem to have been:

(i) Passenger transport—The solution here has been to set up local Passenger services by bus and rail. A national bus company has been get up to take over the bus services previously run by the Transport Holding Company, and with power to acquire voluntarily other under takings.

(ii) Road Policy—The close control of road haulage, which had been a feature of transport policy since the

1930s, was drastically reduced. Directly the Bill became law all vehicles less than 30 cwt. were freed from licence requirements. This meant almost a million tons of capacity became available to carry small loads.

Road transport is particularly suitable for small loads over short distances. The control of larger vehicles, which previously had to e licensed individually, was changed to a system of operators' Licences. This meant that a firm which could rove that it had properly qualified drivers, mechanics, etc., and an adequate financial structure could obtain a licence as a vehicle operator.

It was then free to buy as many vehicles as it liked to operate, and a strongly competitive situation developed, with larger firms able to develop a full service to customers without objections from small operators.

(iii) Railway Policy—The railways were thus forced to face quite severe competition from an increasingly efficient road haulage industry, but were compensated to some extent by being set free from the accumulated debts that had developed over the years since nationalization. Fuller use of railway workshops and vehicle-maintenance departments was made by permitting these organizations to compete with private companies in the sale, manufacturer and repair of a wide range of goods.

The official view her is that is ridiculous to have these expensive assets which have been provided at public expense, under-utilized. Since 1975 the Government has provided large sums to assist the transfer of goods from road to rail to subsidize uneconomic but socially desirable lines.

There has also been in the last few years an agreeable

amount of railway enterprise shown in chasing new sources of revenue, with special terms for pensioners, family travel, etc.

As the United Kingdom transport system enters the 1980s the 44 tonne lorry has been sanctioned, the motorway system is almost complete, and some residual cuts in the rail network are being proposed.

Road Transport—A Growth Industry

There are 200,000 miles of roadway in Great Britain, of which about 9,000 miles are trunk roads, 80,000 miles are classified roads, nearly 1,800 miles are motorways and the rest are minor roads.

In 1980 the railways carried 18,000 million tonne-kilometres—in other words, more than four times as much. 70 percent of rail traffic was coal, coke, iron and steel. The 30 percent of general cargo represents only about 7 percent of the general cargo being carried by road haulage. This gives some idea of the importance of road haulage in recent years.

Use of Containers in Transport

Container services were started as early as 1928 by the four British main-line railway companies, but they have now become firmly established in road, rail, and sea transport. A wide range of off-the-peg containers is now available in a variety of materials, and, while standardization of size is important, containers suitable to the personal needs of most private users can be obtained.

The basic advantages of containers are low freight charges and door-to-door security. Where individual parcels, cartons, or crates are sent by a carrier the work involved in loading and unloading many small items makes for expensive use of labour. If a shipper or distributor is moving large enough consignments go fill a container he can load it himself, either by stacking goods on one another or by making use of the shelving and portioning systems devised by the designers.

The container is then sealed, and moved as one unit with the advanced lifting equipment available. It is impossible to tell what is in it, which is a deterrent to thieves, and without lifting tackle it is impossible to spirit the whole consignment away. Freight charges, insurance, and losses due to minor pilfering are reduced. Container design is important, and handling presents problems. If sea-going containers are to be piled six high, which is no uncommon, the containers must be strong enough to take the weight.

This usually requires a steel framework, but steel is corrosive. Stainless steel is better, but is more costly, and stainless-steel framework with aluminium panels is a useful compromise. Bottom-lift handling devices reduce the need for strength in the container. Top-lift handling devices rely on the container itself to support the weight and place considerable strains upon it. It appears that current container rates are working out about 25-30 percent cheaper than ordinary freight charges, and therefore considerable advantages are offered to commercial firms. In 1979, 32 million container units were moved through ports around the world.

Motorways, Bridges and Tunnels

Road transport growth has slowed down since 1973, when increased fuel charges caused firms to improve economic use of road haulage fleets, but an expanding road network is crucial to progress.

The motorway is the answer to some of the problems giving fast journeys between towns. A start on the motorway network in Britain was made in 1958, and considerable progress has been achieved already in linking the main areas of the country. By 198 Britain had about 1,800 miles of motorway, compared with Germany's 4,200 miles, Italy's 3,500 miles and France's 2,600 miles.

The dilemma of transport is revived by these modern highways, for while the cost is borne socially the benefits in cheaper running costs and faster journeys are enjoyed by the individual families and firms who use the motorway. Tolls have been suggested as a way of reducing the social costs of these projects, but vehicle taxation is still the favourite method of paying for motorways.

The effect of motorways on the location of industry and the availability of labour is considerable. Country towns which once had an unemployment problem, and where emigration to the cities has been the solution for two centuries, suddenly find that land values are soaring as access to the motorway makes the town a desirable site for industry of every sort. Workers begin to travel in from farther afield to take employment.

Bridges and tunnels are essential if motorways are to take direct routes across river estuaries, and some splendid work has been done since the War in bridging the Firth of Forth and the Bristol Channel. The drawings of the Seven Bridge, and its position in the motorway network to South Wales illustrate the grandeur and utility of such engineering feats. The Darford-Purfleet tunnel is a similar link across the Thames estuary.

It has recently been duplicated, so that now two one-way tunnels are available. It is being paid for, like the bridges mentioned above, by tolls charged to users. By giving direct access to Kent and Essex from the other side of the river it is promoting the prosperity of both counties.

Rail Transport

For a century the railways were supreme in inland carriage, both for goods and passengers. Today there is hardly a country in the world where the railways do not lose money. In Britain the railways were nationalized in 1974, and a complete overhaul of the railways system has replaced steam power with electric or diesel-electric rolling stock. High degrees of efficiency have been achieved on particular trunk routes, yet the overall system is still losing money.

Advantages and Disadvantages of Rail Transport

(i) Railway transport is dominated by the 'terminal' problem, for heavy capital costs preclude running a line to every customer's door. Goods must therefore be loaded and unloaded at terminals to complete their journey by road vehicles.

Even though sophisticated handling equipment is now available to load and unload trains, delay seems inevitable and demarcation disputes between group of workers are common.

(ii) Rail transport is impersonal, so that pilfering, delay, and negligent behaviour are difficult to locate and remedy. The consignors and owners of goods are powerless themselves to expedite the handling of goods. One Wiltshire firm sending daily supplies of packages perishable foodstuffs to South Coast towns found it could not get its packing cases back in less than an average of three weeks.

This may not have been any fault of the railway, the may have lain with the retailers, but the impersonal nature of the service operated against the supplier and forced him to invest in much larger numbers of containers than would have ben necessary otherwise.

(iii) Traffic density presents a problem to railways, which operate ideally when the train is full. Highly concentrated traffic is the best. Even very dense traffic at very high speeds can be sagely operated with modern equipment. The Inter-City trains run by British Rail commonly reach speeds of 100 miles per hour.

(iv) Rail traffic is fast when actually moving, for it has a private way which is kept clear. It follows that if goods are to go a long way, rail transport is probably best. The economic limit seems to be about 200 miles. Below that limit the disadvantages of rail traffic make road transport more economic.

(v) It is economical in the use of labour, one driver and guard can take 60 carriages or trucks. Compares with road, transport, where long-distance lorries carrying with the train.

(vi) Because of the nature of the way, a breakdown, strike, or go-slow movement holds up all traffic behind so that its effects are cumulative, building up an enormous backlog of goods which may include perishables, live-stock, etc.

Freightliners—The Railway Container Service

In March, 1963 the Beeching Report 'The-Reshaping of' British Railways' proposed that freight thinner trains be introduced. They were to use road and rail transport for door-to-door delivery of 'containerized' goods, the change-over from road to rail being effected by completely mechanical means at special rail terminals.

By 1970 the scheme had developed to the point where a national freight grid had been established between 29 freightliner terminals in main industrial and commercial centres. Four of these are in London, at King's Cross, York Way, Stratford, and Willesden.

Others are a Glasgow, Manchester, Liverpool, Aberdeen, Birmingham, Cardiff, Edinburgh, Hull, Leeds, Newcastle, Sheffield, Stocketon, Belfast, Dublin, Felixst6wn, Harwich, Nottingham, Southampton, Swansea and Tillbury. All these terminals passed into the control of the National Freight corporate under the 1968 Transport Act in an attempt to integrate road and rail transport more effectively.

The experiment was not really successful, and control of the Freightliner system reverted to British Rail in 1978. Despite fierce competition from a thriving road haulage industry the Freightliner system has certainly kept a useful amount of traffic off the highways, and at times of industrial disputes in the road haulage industry it has proved invaluable.

Beeching Report on the Railways

Continuing heavy losses on the railway network led the Government to appoint one of Britain's leading businessmen, Dr. Beeching, to head a full scale inquiry into the railway organization. His report made drastic suggestions for ending the deficit. They included the following:-

(i) Reducing the rail network itself from 17,000 miles to 8,000 miles, eliminating in the process many rural passenger services, suburban services, stopping passenger trains, and country stations.

(ii) Uneconomic freight trains were to be replaced by liner trains running at regular times with a limited number of stopping points. By having fewer but larger depots increased efficiency could be obtained, but at some loss of convenience to the customer.

(iii) Concentrating on the profitable, long-distance, dense traffic routes, chiefly London-North, London-Midlands, London-West-South Wales, and Glasgow-Manchester-Bristol lines. Nearly all the these were carrying 1,000 tons per day in each direction; some were carrying almost 2,000 tons.

In fact, for political reasons, the full Beeching scheme was never introduced and Dr. Beeching himself left the railways system at 11,000 miles, and the Transport Act of 1968 has implemented a modified scheme. The effect of the Beeching Report has still been very great.

Air Transport

The transport of goods and passengers by air is the greater transport achievement of the post-war era. The great advantage transport is its speed, and for this reason it is particularly suitable for the transport of passengers. Shipping lines that once were prosperous, with full bookings on the New York, South Africa, and Australia runs, have had to turn to cruises and the emigrant trade to earn a living.

However, air transport is expensive, since aeroplanes are sophisticated technical products. Design, operation, and maintenance are all costly, and the actual pay load is small. Even the giant air-buses now in service carry only a few hundred passengers.

The se of aircraft as freighters has been slower to develop because the advantages of air freighting have been less obvious. Only when Air Canada began to publish its market research on air-freight costs did the work wake up to the fact that air freighting was a practicable and economically advantageous method of transport for many industrial and commercial enterprises.

Previously it has been thought that it was advantageous only for perishable commodities or small packets of high value, or for freight directly related to passengers. Even so, the cheapest place for cargo is on the passenger aircraft.

The Economic Viability of Air Freighting

The essential point concerning the transport of goods by air is that the actual cost of the air-freight ticket is less importance than the overall cost of moving the goods from their present position to their destination. The overall cost includes may other expenses besides the actual carriage charge, some of these are listed below.

(i) Factory Condition can be preserved. In many trades goods are finished at the factory and sent to the consumer in first-class condition: dresses and suit are pressed; motor bikes are tuned; bicycles are assembled for delivery.

The need to pack and ship goods destroys this sort of activity, which has to be done at the destination, with high costs of establishing depots, employing and training skilled labour, etc. With air freight, special racks installed in the aircraft rises are high, but the duration of the journey is so short that for any given transit the risks are much smaller than by surface vehicle.

(ii) The saving of time may prevent losses by idle factories or overstocked warehouses. One case on which Air Canada reported concerned a leading German hosiery-making firm which ordered 230 machines from Pennsylvania. The air-freight charge was Rs. 2400 per unit, six times more then the cost of dispatch by sea.

The machines were installed ten days ahead of schedule as a result, and turned out 207,999 pairs of hose in the ten days. This was more than enough to yield a handsome profit on the venture. Another firm flew in its raw materials and components from suppliers daily, so that there was no need to warehouse or stockpile these goods.

Its saving in warehouse space alone permitted an extension of plant layout to achieve economies of large scale, which more than off-set the high cost of air freight.

(iii) The cost of packing. Packaging can be very expensive for a sea voyage, where inclement weather may damage the goods salt air may taint produce, humidity may rust metals, etc. With air freighting most packaging is quite unnecessary. Modern aircraft fly above the clouds nowhere near the sea, and apart from possible effects of cold, the goods in transit are unlikely to come into contact with 'weather.'

(iv) Insurance costs are lower by air. Aircraft risks are high, but the duration of the journey is so short that for any given transit the risks are much smaller than by surface vehicle.

(v) Where goods are fashionable, they sell best on the crest of the wave. One manufacturer of wallpaper filled a charter aircraft with rolls showing a popular recording groups and followed them around America selling at fancy prices.

These are some of the aspects which, when taken into account, make air freighting an extremely useful solution to man's transport problems.

Sea Transport

Since the massive development of the air transport industry the passenger trade for long-distance sea voyages has almost disappeared, and the types of vessel in sue have changed. The chief ones are:-

(i) Container Ships—These are a ships which carry cargo in containers, rather than in conventional holds. The cellular construction permits containers to be loaded into the holds. The cellular construction permits containers to be loaded into the holds quickly, the actual 'stuffing' of the containers to be loaded into the holds quickly, the actual 'stuffing' of the containers having been done on shore.

The containers may be stacked up to seven high, and an entire ship can be turned round in twelve hours.

(ii) Bulk Carriers—A ship built for a particular type of cargo, e.g. iron ore, is often called a bulk carrier. The emphasis these days is on larger and larger bulk carriers, because the cost per ton of cargo is reduced.

(iii) Oil Tankers—In the last forty years the tanker had become the commonest ship on the high seas, and a wide variety of them is available. The largest are the V.I.C.C's which operate from the oil-producing countries of the Middle East, Nigeria, Venezuela, etc., to the refineries of Europe, North America and Japan.

(iv) L.N.G—Carriers are special refrigerated tankers for liquefied natural gas, while L.P.G. carriers carry liquefied petroleum gas. Clean product carriers, often quite tiny, carry refined products from oil refineries, chemical works etc. Bulk transport of liquid products is cheap and convenient. In former times oil, for example, was moved in barrels, which had to be lifted on and off the ship by cranes.

(v) Roll-on roll-off Ferries—There are great advantages in sending goods by road on long-haul journeys. The roll-on roll-off ship enables lorries and cars to drive onto the ship to cross from one land mass to another. Even the transport of cars from continent to continent is economical by this method.

(vi) Passenger Liners—As explained, the huge passenger liner has largely disappeared from the seas except as a cruises liner for holidays at sea. There are passenger cargo liners which are chiefly concerned with cargo to and from the major ports of the hull is taken up with the holds of the vessel.

(vii) Lash Ships—'Lash' stands for 'lighter aboard ship.' These huge ships carry barges from fiver mouth to river mouth, collecting loaded barges at the mouth of the Mississippi and releasing them to journey from Rotterdam up the Rhine to Germany, or up the Thames to inland wharves. Bacat stands for 'Barge-aboard-catamaran,' a similar type of vessel to a lash ship.

(viii) Tramps are pure cargo ships which travel anywhere in the world to earn a living. Like a tramp they have no steady home. A typical journey might be as follow:-

(a) Form Australia with wheat to China.

(b) Japan to West Africa with manufactured goods and electrical engineering equipment.

(c) Cargo of motor vehicles and engineering parts for South Africa.

(d) Across to India for a cargo of cotton goods for Australia.

(e) From China to Japan with iron ore.

(f) Home to the United Kingdom with cocoa and palm oil.

Terminals for Ships

Passengers and cargo have to have access to ships. The terminal is the port, where ships can be serviced, provided with all their requirements before the voyage, and take on passengers and cargo.

The Importance of Shipping

Britain is an island inhabited by over 50 million people. We could not possibly grow sufficient food to feed our vast population ourselves and we have insufficient supplies of most raw materials. We therefore need to import food and raw materials of every kind. In order to pay for these we must export all sort of finished goods and services. These quantities of goods can be moved only by sea, for air transport is too expensive for heavy goods. Shipping is our most important link with foreign parts, and on it depends the prosperity of our people.

Reference is often made to liner Conferences. The conference system is a way of protecting the liner trade from unfair competition from tramp shipping at times when there is excess capacity in the tramp market. If liners are to provide regular services calling at scheduled ports and leaving at stated times irrespective of the cargo position, they must be assured of reasonable freight rages. Tramp freight rates fluctuate with supply and demand, but liner freight rates are held steady. At times when tramp rates are high the liners carry at their steady rates.

When tramp rates are low there is a temptation for liner customers to transfer their cargoes to the cheaper tramps. If this happens the liners will suffer losses and may be driven out of business. To encourage regular support from traders, the conference system offers rebates which can be enjoyed only if the trader can prove that his goods have on every occasion in the previous trading period travelled by the conference liners.

This conference 'tie' permits a trader to ship his goods by any line that is a member of the Shipping Conference. These liners charger similar freights, but it is unfair to regard them as price-fixing 'rings' since they have in no way a monopoly of shipping, which is international in character and highly competitive.

The International Chamber of Commerce, which represents the liners' customers, itself approves the conference system because of the security and reliability of service that the liner owners give to businessmen.

Requirements of a Good Port

(i) A clear channel—This may mean dredging, or blasting rocks out of the may, or marking the deep water with buoys and lights. It will certainly be necessary to avoid blocking the channel with sunken ships. Pilots who know the way into the harbour are given sole charger of vessels entering and leaving.

(ii) Government offices for customers control, emigrating, immigration, licensing, quarantine centres, police and security establishments, public health and sanitation.

(iii) Deep water—This may be permanent deep water or, if it is tidal water, the water in the dock must be kept in by lock gates when the tide goes out. Southampton is a port with a natural advantage; it gets four tides a day.

(iv) Shelter—This may be in a natural bay or harbour, or it may be provided by the banks of a river. If these are not available groynes must be built.

(v) Supplies of coal oil, water, food, and electricity for ships calling at the pot. When a ship arrives it will need to re-fuel and replenish its stores of water and food. While its engines are not working it may need to use the shore supply of electricity.

(vi) Wharves, Cranes, floating cranes, and brags are needed to help unload the ship, and turn it round quickly. A vessel is wasted in port—its function is to sail the seas. A labour force of coders and stevedores is needed to carry out the work of loading and unloading.

(vii) Commercial Premises—warehouse, bonded warehouses, refrigerator stores, grain elevators, forwarding depots for packaging exports, insurance offices.

(viii) Dry docks, repair yards, workshops, disinfestation equipment and maintenance centres.

The Chief Ports of the British Isles

The map above shows the ports of the British Isles, and the shipping lanes which terminate in them. The importance of any port depends to some extent on its position and the industries and population it caters for.

1. London—London, under its Port of London Authority, is still wrestling with the problems of adapting to Britain's changed position in the world, and the technological changes in shipping since containerization began in the sixties. Once the busiest port in the world, its up-river docks are now inaccessible to the supertakers and bulk carriers; its passengers trade is dead except for a tiny cruise trade and the economics of the container trade favour a major stopping place in Europe for most Rhine for bare traffic and its absence of serious tidal problems, is a major competitor. P.L.A. plans to build a major deep-water port at Maplin near Shoeburyenees have been frustrated by official policies, but are not quite dead. It is a major container port. Far enough down river to accommodate large ships, it has been chosen as Britain major container terminal.

The facilities when have come into use are some of the most advanced in the world. Six Paceco-Vickers container cranes straddle carriers and other handling equipment provides rapid turnaround for container ships.

The British container consortium O.C.L. uses Tilbury for its Australia service. There are roll-on-roll baths, a common-user berth and also two prepackaged timber berths for the handling of wood products, manned by a permanent crew of skilled dockers who have set up new handling records with their side-loader fork-lift trucks. Just outside the container port in the main river a new grain terminal has been completed which can turn round large bulk carriers in two days. Land has been developed by the creation of modern mills and grain-processing plants and plans are in hand for four deep-draught riverside berths.

2. South-East Ports and Trade with the E.E.C— Two fifths of all U.K. exports and imports now go to, or come from, the South-East ports, While London dominates the South-East, its port is less well placed, partly because of the tidal nature of the River Thames, but also because of excess costs imposed by a traditional dock labour force.

Other ports have been able to expand more easily, with less sophisticated labour, shorter sea routes and absences of tidal problems. Southampton Dover, Harwich and Felixstown have picked up the profitable short-sea container trade at London's expense. They have all expanded threefold in the decade.

Milford Haven is Britain's second largest port in terms of tonnage handled, but nearly all its cargo is petroleum products.

Immingham, across the Humber Estuary from Hull, is now the third largest port, also handling mainly bulk cargoes.

Southampton, the next most important port of the

British Isles, has a residue of passenger trade with cruise ships taking a larger share of this business in recent years. It has become an important container port too, adding this type of packaged cargo to its traditional high-speed cargoes of refrigerated meat, fruit and vegetables.

Liverpool has declined in influence as U.K. trade has become increasingly Europe-orientated. Its rise to power began with the cotton and tobacco trade about 250 years ago, but its growth has been hampered recently by the decline in foreign and non-European trade.

Hull, the convenient port for the industrial cities of the West Ridding, is another important British port. Its fish trade has declined, but it imports wool from Australia, wheat, and a wide variety of general cargo.

Glasgow grew to importance when the New World was opened up in the sixteenth century, handling in particular the tobacco trade. Later, the fortunate discovery in the Glasgow area of coal and iron ore led to the growth of heavy industries and a large export trade.

Designated as Britain's first M.I.D.A., the Clyde Port Authority has deep water suitable for large bulk carriers, but is on the wrong side of the U.K. remote from Europe. The For the part Authority, which includes Leith and Grangemouth, is the port authority for East Scotland. It handles cargoes from Scandinavia, the Baltic and the Low Countries.

Bristol, a port that once gave its name to the British sailor's idea of efficiency—'ship-shape and Bristol fashion'—has declined because it is too shallow for modern shipping. Instead the port of Avonmouth was built nearer to the sea by the Bristol authorities, and a recent addition has been a now dock, the Royal Portbury

Dock.

Other ports. Several small ports have demonstrated since the Second World War that the smaller terminal has much to offer the carrier. Outstanding among the small ports are Flisxstowe, which developed a container berth before the bigger ports had realized the revolution that was to come over shipping. Middlesbrough now handles 16 million tonnes per year and the Midway Ports, in Kent, 11 million tonnes.

Advantages and Disadvantages of Sea Transport

(i) The 'way' is free, and gives access to most parts of the world.

(ii) Small power is needed to drive a vessel, so that large vessels are practicable and economies can be achieved.

(iii) Transport by sea is relatively slow, so that passengers are easily lost to airlines when speed is important.

(iv) The possibility of determination of cargo is great because sea water is corrosive and also liable to affect the quality and flavour of cargoes.

(v) Because weight is not important, ships can be massive enough to be strong.

(vi) Buoyancy of the effective weight zero. Hence sea transport is very suitable for heavy goods.

(vii) Pilfering of cargoes is not uncommon, though 'containerization' is doing much to reduce this.

(viii) Ports are expensive and delays not uncommon.

Other Transportation Systems

Hovercraft. Hovercraft operate on a cushion of air which is pumped under the vehicle and kept in by a skirt of heavy material. They are particularly effective over water, where they enable much higher speeds to be achieved than by the conventional ship. There is no water resistance worth speaking of, and a cross-channel ferry is now operating from a base at Pegwell Bay near Dover, carrying vehicles and passengers at about 80 knots.

The hovercraft is particularly advantageous in estuaries where its ability to ignore sandbanks and shallow water means it can operate by more direct routes. The use of longer skirts, possibly up to 12 feet deep will improve the ability of these craft to negotiate the larger waves. A second use for hover skirts has been found in moving heavy loads over bridges which are not strong enough to carry the load. By fitting a temporary skirt to lorries carrying heavy loads of this type the weight which is usually borne by the axle assembled, and enabling the load to cross without damage to the bridge.

Pipe lines are increasing in importance for the transportation of natural gas and oil field from the ports to inland depots. While the capital cost is great, the savings on motor-vehicle costs for comparable movements by road, quite apart from the social costs of congested highways, make this form of transport both economical and socially desirable. Pipelines cannot however compete with tanker costs, especially giant tankers on long hauls.

12

PROBLEM OF ENVIRONMENT POLLUTION

Environment

On the earth, various kinds of living creatures lead their life. Various kinds of things surround them. Whatever surrounds these creatures is collectively known as environment. In other words, all the external things or situations in which human beings are forced to live is known as their environment. This kind of environment is known as external environment. There is another kind of environment also, which is known as internal environment which rests inside the body and mind of the individual. Between these two kinds of environment, a balance is always kept by body however because of various factors or reasons, this balanced state can be disturbed. Out of various factors which can bring a kind of imbalance between external and inner environment is environment pollution.

By living in a polluted environment, an individual can get affected by various kinds of diseases. Whenever any foreign substance enters in the environment, it gets polluted and affects the life of human beings or other living creatures in different ways. Because of this foreign substance, its various constituents, namely, the air, water etc. also gets adversely affected and they also get polluted. In this chapter, we will discuss about various kinds of pollutions which is being found in the constituents of environment, the reason of such pollution and the consequences of their polluted states.

Water Pollution

An important constituent of environment is water and when water gets polluted in an way, it is known as water pollution. Whenever any foreign substance, which can be organic, inorganic, biological or some physical substance, enters in the water, the quality of the water gets deteriorated and this kind of water does not remain fit to use. Polluted water is not unfit for human beings, but also for the other living creatures. The qualities of polluted water get changed. For various purposes, water is being used by human beings in different functions, namely, for quenching the thirst, cooking food, agricultural purposes and washing clothes and utensils.

Polluted water consists of various kinds of impurities and when this kind of water is being used by human being in any of the functions, person or involved persons gets affected to some extent. By using polluted water, various kinds of intestinal infections can take place in human beings, because of which it is always better to avoid using this kind of water.

There are various reasons which led to pollution of water. It is because of various activities perform by human beings that water gets polluted. Two important factors are urbanisation and industrialisation. Today, we see that various people have migrated in big cities from the small villages because of which cities have been over-crowded. Various factors have been established for different purposes because of which water gets polluted.

The main activities which led to the pollution of water are as follows :—

i. Industrial Wastage : Various kinds of products are being prepared in the factories, in the production of which various waste products emerge. These waste are highly chemical products consisting of complex synthetic organs. Water gets polluted when the waste products originating in the factories get mixed with it.

Almost in all the factories, provision is made that waste products are being thrown in the water river and when these products get assimilated in water, various kinds of impurities in different forms enter the water and make it imperfect to use. When these chemicals are absorbed by the sand or the earth, the quality of underground water gets deteriorated. In big industrial cities, this kind of problem is increasing day by day. The substance from manure factories, the waste material from the mills preparing the chemicals render water impure. By living in impurer water, health of the living creatures living beneath the water also gets affected.

ii. Sewage : An important source of pollution in urban cities is sewage, which consist of decomposable organic matter and pathogenic agents. In developing countries, people are migrating in a large number to the big cities, because of which sewage is being produced in large quantity, but disposal of these products cannot take place accordingly. When the impure liquid get merged in the water, or they are being stored in any open space, after some time, mosquitoes breed over them and various kinds of bacterise and impurities enter into the water, which makes the water impure. With this, oxygen contents in the water get depleted.

iii. Physical Polluteness : Thermal pollution and radio-active substances are the main constituents of physical polluteness. Therma pollution takes place because of hot water. It is due to this hot water that the temperature of aquatic environment get increased. With this, aquatic animals and plants get affected adversely. When radio-active ash falls into the water, it gets polluted, which result from processing of uranium or research laboratories where isotopes are being used. On coastal areas of lakes, seas and oceans, oil gets spill. Oceanic water also gets polluted. Marine pollution results by the mineral oil which is being discharged from ships. By all these ways, water of ocean or sea gets polluted, which affects the fish or living creatures leading their life under the water. When these fishes are being eaten by human beings, their health also get affected adversely.

iv. Agricultural Activities : Today our country is developing or progressing in all the fields and this kind of development can be seen in the agricultural sector also. Today, primitive means of agriculture are not used, but various kinds of advanced equipments or methods are being used. To protect fields from insects, various kinds of chemicals are being used. If these chemicals or fertilizers are being used in greater quantity, they affect the quality of land adversely. When these fertilizers get dissolved into water, they remove the hydrogen present in water, which affect the life of various living creatures leading their life under water. When this kind of polluted water is being taken by human beings, their health gets deteriorated and they get affected by various diseases.

Thus we see that there are various causes which lead

to the water pollution. The main reason which emerge from the above discussion seems to be increasing population and the increasing migration from villages to towns.

Today, water pollution has become one of a big hazard to the health of the people. Various kinds of diseases emerged out by using polluted water. All the living creatures become affected by polluted water directly as well as indirectly.

Various measures or steps have been taken by the government to clean the water of various rivers and in making this kind of polluted water clean, large amount of money is being spent every year. Thus, water pollution leads to economic loss to the human beings indirectly.

Water pollution is the consequences of use of various modern industrial technologies and increased population. Now the problem which is being faced is to determine the optimum level which can permit economic and social development of the people, without affecting health in any way. For controlling or putting the stoppage on the water pollution, various measures are being used by the government, some of which are as follows :—

a. Population Control : As mentioned earlier that one of an important cause or reason of water pollution is increase in the population. Various programmes or policies have been designed by the government of India to control growth rate. Because of the tremendous population, people do not get the opportunity to get educated and in case of illiteracy, it becomes very difficult to convince them that they are responsible for water pollution. For this purpose, government should educate the people as well as try to control the population of the country.

b. Making waste water Clean : Industries and tremendous increase in the urban population are considered to be the two main causes of water pollution. Industrial waste is known as effluent, while the urban liquid waste is being called sewage.

Sewage is being produced by domestic as well as from the industrial area. This kind of product is being produced from various human activities. Sewage consists of pathogenic germs, and is very harmful and hazardous for the health. Sewage should nøt be thrown into water body without being treated properly. By their treatment, we mean extracting impurities from them. When sewages are being treated properly, the chances of spreading the communicable diseases from them gets reduced. Collection, treatment and re-use are the main steps which are being included in the treatment process of sewage. All the pathogenic agents preset in the sewage are being destroyed through the various treatment processes. From treatment processes, sewage becomes suitable for re-use.

Another measure which is being used by government of India is use of various methods in relation to industrial waste water treatment. For this, three steps are being identified, under which, various kinds of pollutants are being recognized. Pollutants are considered to be of three different kinds, namely, the floating pollutants, suspended pollutants and dissolved pollutants. Oil, grease and fat are the main kinds of floating pollutants. These kind of pollutants generally affect the growth of the plants. Suspended pollutants can be either organic or inorganic in nature. Acids, alkalis, heavy metals and insecticides are known as the dissolved pollutants, which can affect the aquatic life adversely.

Dissolved air, floating, chemical treatment, biological treatment are main methods with the help of which industrial waste water is being treated. If water is being treated properly, it can be re-used in various functions which can also solve up the problem of water shortage in big cities.

iii. Legislation : Water of ocean, rivers, canals get polluted in various ways because of which problem of polluted water has become a globalised one. To protect the water from being getting polluted, some important steps have been taken by the government. In 1974, parliament passed the water act, the reason of which was increase in quantity of polluted water. For controlling pollution, this act is provided with constitution wide powers. It is through this legislation that it was being set out that no industry is permitted to establish without furnishing the certificate by the owner in regard to re-treatment of waste water of the industry.

Method to Purifying Surface Water

As mentioned earlier that water is one of an important constituent of human body and water of canals, rivers and lakes is mostly used for drinking, but water in these sources do not remain clean or purified. Before consuming this kind of water, it is necessary to make them fit to drink. There exists various kinds of impurities in this water which make them unfit for the people to drink. With the help of some special kind of processes, water can be purified, some of the methods being used for this purpose are as follows :—

a. Decantation : In this process, impure water is being

pumped and collected in a decanting basin when soil and other waste products get settled down after some period of time.

b. Filtration : From settling basin, water is being transferred from filter, which consists of sand in some quantity. With this process, almost maximum impurities lying in the water gets destroyed. It is necessary to clean the filter regularly.

iii. Chlorination : After bringing the water out of filter, the next step is to mix chlorine in it. With this method, bacterias present in the water can be removed.

Air Pollution

Air cannot remain pure for a long period of time. For all the living creatures, air is the important and one of the closest component of environment. Oxygen for respiration is being provided by the air. It is through the air present in the atmosphere that the various kinds of sounds and smells are being transmitted from one place to another. When the various hazardous elements get mixed in the air, air becomes polluted. When foreign matter in the air get concentrated in a large quantity, air is called to have been polluted. Throughout the world, the problem of air pollution is becoming more serious.

Various experts have devoted their time in analysing the problem of air pollution, and it has been found by them that there exist some specific causes or sources which can pollute the air, some of which are as follows:—

i. Combustion Sources : By combustion of various fuels, various harmful smoke gets mixed in the environment which also make the air dirty. In the combustion sources, dust from domestic as well as from industrial sector are being included.

ii. Industrial Processes : Various kinds of processes are being performed in the industries and these processes result in various polluted gases and dust. When these gases merged with the air, they make air polluted.

iii. Motor Vehicles : Today, our life has become mechanized and for moving from one place to another, we make use of various kinds of motor vehicles. These vehicles are run through motors and they produce smoke which is very impure. Upto a normal limit, this kind of smoke emerges from the engines, but when there is any defect or problem in the engine, the amount of dust increases, which makes air polluted.

Although through the world, problem of air pollution has taken a serious face, but people of under-developed countries have to face them utmost. In such countries, processes of industrialisation and urbanisation are done without any kind of planning. Various air pollutants have been chosen and all these are considered as indicators of air pollution. Sulphur dioxide, carbon-oxide, and nitrogen are some of the important pollutants. Firstly, it is measured that in what quantity these substances are present in the air and if they exist in excess to the permissible limit, the air is considered to be polluted.

Consequences of Air Pollution

By living or breathing in the polluted air, health of the living creates get affected to a lot of extent. By air pollution, respiratory system gets affected to considerable limit. Chronic bronchitis is the disease which generally affects the people who breath in polluted air for longer period of time.

In the polluted air, plants and animals do not get grow to their maximum capacities. Plants grow improperly and the leaves of these plants get burnt or destroyed within short period of time. By inhaling in polluted air, animals suffer from various respiratory diseases.

Air pollution is effecting ozone layer adversely, because of which percentage of ultra-violet rays on the earth has increased, because of which plants and animals are being affected in various ways. By pollution, property and material are also affected considerably. Pollutants of the air pollution are so strong that they can corrode various metals, stones and marbles. Rubber can be cracked by the ozone depletion.

Social and economic aspects of an individual get also affected by the air pollution. Visibility of the people have affected adversely because of increase in the air pollution.

Important Ways to Control Air Pollution

For the prevention of air pollution, various steps have been taken by various international societies, and one such society is World Health Organization, which have done an enormous step in 1968 by publishing a document entitled Research into Environment Pollution, recommending certain measures with the help of which air pollution can be controlled, which are as follows :—

a. Replacement : In this measure, those products or processes which can result in air pollution are being replaced by other products and processes which do not create any kind of pollution. It is according to this concept that government of Delhi has introduced the concept of CNG instead of diesel buses.

b. Containment : Provision of various enclosures, ventilation and air cleaning are introduced in this process, the main object of which is to stop the release of toxic substance into the air. Various engineering methods can be used for this process.

c. Legislation : For controlling the problem of air pollution, various legislations have been adopted by various countries of the world. It is according to this process that government of India has passed Nuisance Act in our country.

iv. Dilution : By planting more and more plants, pollutants are being controlled under this method. An attempt in this regard is to establish green belts between industrial and residential areas. This means or mechanism proves to be fail when the amount of pollutants increase in the air.

Concept of Ozone Layer

The air in which we breath is composed of various gases, two of these main gases are nitrogen and oxygen. In addition to these, carbon dioxide, hydrogen and ozone are some of the important constituents of the air. It is believed by some people that the refreshing air we breathe in at the sea-side is known as ozone, however, the truth is that ozone is a poisonous kind of oxygen. Very little ozone is being found at the ground level. Around 24 km above the earth, there exist a very thin layer of this gas. Ultraviolet rays emitted by the sun are being blocked out by this ozone layer. Ultraviolet rays are very harmful for the living creatures thus, this layer prevents this rays to reach on the earth.

Because of the increasing pollution, this invaluable layer is destroying day by day. Ozone layer is getting depleted with the increase in air pollution. It has been found by the experts that there exists a large hole on the north pole of the layer. For saving the ozone layer from deterioration, it is becoming very necessary to control the air pollution with any means.

With the depletion of the ozone layer, all the living creatures will be affected adversely. All the human beings then have the small sized eyes. They will have long nose and small height.

Noise Pollution

Another problem related to the polluted environment is noise pollution. Generally, this kind of pollution is found in big cities or in the over-crowded places. Unwanted sound is known as noise. When the wrong sound is being produced in the wrong place and at the wrong time, it is called the noise. The condition in which excessive sound is being forced into the atmosphere without any use or necessity, it is called the noise pollution.

Fundamental Causes of Noise Pollution

There are various sources which can give rise to the noise in one way or the other. All the sounds emerging from household and industrial areas are included in it. Importance characteristic of noise, which differentiate it from the normal sounds are as follows :—

a. Noise has a high level of loudness. It is on the amptitude of the sound waves produced from the source that the intensity of the sound depends. The measurement unit of noise is decibel. Human ear can hear one decibel, which is the small amptitude.

b. Noise has high frequency, which is being denoted by Hertz.

To measure noise level frequency, duration and other environmental factors which are being used in industrial places, various kinds of instruments are being used.

Consequences of Noise Pollution

Unwanted noise affects the health of human beings adversely. An individual can be affected by noise pollution in various means. Simple effects of noise pollution are annoyance, irritation and disturbing sleep. If someone remains constantly in an environment in which noise has very high level, various kinds of serious diseases or effects can be found in him. Hearing power of such individual can be impaired and after a long period of time, he can become deaf. Various kinds of psychological and physiological changes can result in human beings because of noise pollution. Researchers has found a direct relationship between the levels of noise pollution and heart diseases. Efficiency to perform various functions can be affected by noise pollution. Behaviour and habits also get changed after some time living in an environment affected by noise pollution.

Controlling the Noise Pollution

While adopting remedial measures, some aspects should be provided the consideration, which includes the source of noise, medium of noise and the object being affected by noise.

To control the noise pollution, people should be made aware of the reason and consequences of noise pollution, for which various kinds of means or mediums should be used. Steps should be taken to put a kind of control on the source of the noise. Factories, in which noise machines are being used should locate at far distance from the residential areas. The limit should be imposed on the level to which loudspeakers can be used and time duration should also be defined by the government. If any vehicle is producing greater sound, the user can be asked to send the vehicle for service, or a penalty can be imposed on him. Horns of the vehicles should not be very loud.